AF700423

Agent-Based Social Systems

Volume 9

Editor in Chief:

Hiroshi Deguchi, Yokohama, Japan

Series Editors:

Shu-Heng Chen, Taipei, Taiwan, ROC
Claudio Cioffi-Revilla, Fairfax, USA
Nigel Gilbert, Guildford, UK
Hajime Kita, Kyoto, Japan
Takao Terano, Yokohama, Japan

For further volumes:
www.springer.com/series/7188

ABSS–Agent-Based Social Systems

This series is intended to further the creation of the science of agent-based social systems, a field that is establishing itself as a transdisciplinary and cross-cultural science. The series will cover a broad spectrum of sciences, such as social systems theory, sociology, business administration, management information science, organization science, computational mathematical organization theory, economics, evolutionary economics, international political science, jurisprudence, policy science, socioinformation studies, cognitive science, artificial intelligence, complex adaptive systems theory, philosophy of science, and other related disciplines.

The series will provide a systematic study of the various new cross-cultural arenas of the human sciences. Such an approach has been successfully tried several times in the history of the modern science of humanities and systems and has helped to create such important conceptual frameworks and theories as cybernetics, synergetics, general systems theory, cognitive science, and complex adaptive systems. We want to create a conceptual framework and design theory for socioeconomic systems of the twenty-first century in a cross-cultural and transdisciplinary context. For this purpose we plan to take an agent-based approach. Developed over the last decade, agent-based modeling is a new trend within the social sciences and is a child of the modern sciences of humanities and systems. In this series the term "agent-based" is used across a broad spectrum that includes not only the classical usage of the normative and rational agent but also an interpretive and subjective agent. We seek the antinomy of the macro and micro, subjective and rational, functional and structural, bottom-up and top-down, global and local, and structure and agency within the social sciences. Agent-based modeling includes both sides of these opposites. "Agent" is our grounding for modeling; simulation, theory, and real-world grounding are also required.

As an approach, agent-based simulation is an important tool for the new experimental fields of the social sciences; it can be used to provide explanations and decision support for real-world problems, and its theories include both conceptual and mathematical ones. A conceptual approach is vital for creating new frameworks of the worldview, and the mathematical approach is essential to clarify the logical structure of any new framework or model. Exploration of several different ways of real-world grounding is required for this approach. Other issues to be considered in the series include the systems design of this century's global and local socioeconomic systems.

Editor in Chief
Hiroshi Deguchi
Chief of Center for Agent-Based Social Systems Sciences (CABSSS)
Tokyo Institute of Technology
4259 Nagatsuta-cho, Midori-ku, Yokohama 226-8502, Japan

Series Editors
Shu-Heng Chen, Taipei, Taiwan, ROC
Claudio Cioffi-Revilla, Fairfax, USA
Nigel Gilbert, Guildford, UK
Hajime Kita, Kyoto, Japan
Takao Terano, Yokohama, Japan

Koen H. van Dam • Igor Nikolic • Zofia Lukszo
Editors

Agent-Based Modelling of Socio-Technical Systems

Springer

Editors
Koen H. van Dam
Faculty of Technology, Policy
and Management
Delft University of Technology
Delft, the Netherlands

Zofia Lukszo
Faculty of Technology, Policy
and Management
Delft University of Technology
Delft, the Netherlands

Igor Nikolic
Faculty of Technology, Policy
and Management
Delft University of Technology
Delft, the Netherlands

ISSN 1861-0803 Agent-Based Social Systems
ISBN 978-94-007-4932-0 ISBN 978-94-007-4933-7 (eBook)
DOI 10.1007/978-94-007-4933-7
Springer Dordrecht Heidelberg New York London

Library of Congress Control Number: 2012948330

© Springer Science+Business Media Dordrecht 2013
This work is subject to copyright. All rights are reserved by the Publisher, whether the whole or part of the material is concerned, specifically the rights of translation, reprinting, reuse of illustrations, recitation, broadcasting, reproduction on microfilms or in any other physical way, and transmission or information storage and retrieval, electronic adaptation, computer software, or by similar or dissimilar methodology now known or hereafter developed. Exempted from this legal reservation are brief excerpts in connection with reviews or scholarly analysis or material supplied specifically for the purpose of being entered and executed on a computer system, for exclusive use by the purchaser of the work. Duplication of this publication or parts thereof is permitted only under the provisions of the Copyright Law of the Publisher's location, in its current version, and permission for use must always be obtained from Springer. Permissions for use may be obtained through RightsLink at the Copyright Clearance Center. Violations are liable to prosecution under the respective Copyright Law.
The use of general descriptive names, registered names, trademarks, service marks, etc. in this publication does not imply, even in the absence of a specific statement, that such names are exempt from the relevant protective laws and regulations and therefore free for general use.
While the advice and information in this book are believed to be true and accurate at the date of publication, neither the authors nor the editors nor the publisher can accept any legal responsibility for any errors or omissions that may be made. The publisher makes no warranty, express or implied, with respect to the material contained herein.

Printed on acid-free paper

Springer is part of Springer Science+Business Media (www.springer.com)

Foreword

The first tentative efforts at 'simulating societies' using agent-based models were made in the early 1990s. Since then, there has been an explosive growth in the application of agent-based modelling in the social sciences, with applications in nearly the whole suite of disciplines, including economics, sociology, geography, political science, anthropology, linguistics and even social history. At first, the emphasis was mainly on small, simple models that could illuminate basic social science theories. For example, we saw models showing how cultures could differentiate, residential segregation could emerge from household moving decisions, political opinions could diverge towards the extremes and other similar examples. In addition, there was a cross-fertilisation with game theory, since agent-based modelling enabled scholars to break free of the constraints imposed by the need to obtain analytical solutions from mathematical formulations.

It has only been within the last decade that the state of the art has advanced to the point where it has become possible to enlarge the scope of agent-based modelling to encompass more applied and policy-oriented topics. This was a result of a gradually improving understanding of the strengths and limitations of agent-based models, together with huge improvements in the supporting infrastructure: much better development environments and much faster computers, which made larger and more complicated models feasible to develop and test.

The team who wrote this book have been pioneers in this move to more applied models. Over the last decade, they have been accumulating experience about how to build policy-oriented models, and what such models can and, as important, what they cannot be used for. This experience is the basis for this book. The book is an immensely valuable resource because it draws on what the team have learned from a long series of applications of agent-based modelling to the understanding of policy options for infrastructure systems and services. Although the team's focus is on infrastructure (such as electricity and transport), these lessons can easily be transposed to other areas, such as health care, education, flood defence, the provision of renewable energy sources—all of which have been the topic of agent-based modelling efforts in the last year or two—and many others.

Agent-based modelling has a number of major advantages as a support for policy making. First, the basic idea is accessible and easy to grasp, even for those who are unfamiliar with the approach. As its name implies, an agent-based model consists of a number of 'agents' represented in a computer program. Each agent corresponds to a real person (or organisation, firm, department or other group) in the real world. These agents are programmed to interact in the same ways as the real actors do and to experience the same constraints and have access to the same knowledge. This one-to-one correspondence between what we see in the policy world and what is represented in the model makes it easy to grasp what the agent-based modelling approach is about, especially in these days of computer games (which are often rather similar to agent-based models, other than that they are designed to be fun, rather than to be good simulations).

A good agent-based model can be relatively 'transparent' to inspection by decision makers. For example, one can try out what happens in the model when agents are given certain attributes, or are allocated particular behavioural rules, and can check that the model behaves in a plausible fashion. If data are available, then the model outcomes can be compared with what actually happened.

A third advantage of agent-based modelling is that it can deal with complexity. It is increasingly being realised that the social world has to be understood as a complex adaptive system, meaning that the interactions between its parts are non-linear and multi-level. These aspects can be simulated in a natural way with agent-based models. Moreover, as the authors point out, when one is dealing with questions of infrastructure, one has to contend not with just a social system, nor with just a technical or physical system, but also with the complex interactions between these two.

Because both the social and the technical have an influence, this kind of modelling requires a disciplinary breadth, as is well illustrated in the pages of this book. One needs to understand and model a host of issues such as consumer behaviour, political and organisational stresses, the operation of markets, the technicalities of the design and manufacture of electronic products, and environmental sustainability, to list just some of the areas covered by the case studies in this book.

This need to be a jack-of-all-trades, as well as having skills in modelling, is one of the challenges of this style of policy-relevant simulation. Another is to avoid 'overselling' one's model. As the authors wisely note, the interests of those who want to use models and those who build models are rarely completely aligned. Modellers are usually acutely aware of the limitations of their models and the way in which their behaviour depends on the assumptions that have been made. In contrast, those who use models to guide policy decisions would prefer simple, definite and precise predictions of what the future will bring. There is thus a constant temptation for modellers to pander to the desires of their 'clients', and for decision makers to over-interpret the advice the model provides. If agent-based modelling in the policy arena is to avoid gaining a poor reputation, it needs to navigate the difficulty territory between offering over-simple answers and being thought to be over-complicated and unable to make any worthwhile contribution.

One refuge in which modellers have often sought shelter is to insist that their models are only able to help with understanding the underlying mechanisms and cannot make predictions about the future. The distinction between understanding and prediction is however a difficult and unrewarding one: while it is rarely possible to make point forecasts about the future behaviour of any complex system, even developing a model for the purpose of understanding involves making some predictions. What modellers do need to do, however, is to comprehend the limits of prediction— for instance, while they may not be able to say what will happen, it is often possible to say with relative certainty what will *not* happen.

The authors emphasise the importance of involving 'stakeholders' throughout the model building process. This helps to keep the model on track, but it also means that the stakeholders will have a stake in the model's design, and will be less likely to dismiss it as the creation of ignorant outsiders. The advice to involve stakeholders at all points in developing a model is just one of the many invaluable hints that the authors sprinkle through the book.

One direction that the development of policy-oriented modelling seems likely to take is to involve non-modellers more and more closely in the design process. Perhaps one day there will be an agent-based modelling tool as easy to use as the spreadsheet, but even if that happens, there will still be a need for a clear and well-tested process for conceptualising, designing and validating agent-based models of socio-technical systems. This book shows what that might look like.

Guildford, UK Prof. Nigel Gilbert

Preface

The book you have in front of you is the synthesised result of several research projects executed at the Energy and Industry (E&I) group, part of the faculty of Technology, Policy and Management of the Delft University of Technology in the Netherlands. Since 2004, the E&I group uses the agent-based paradigm to develop socio-technical models of infrastructure systems and services and their evolution. The researchers soon realised that we could be more efficient by working together and building on top of each other's models, but also that the reality of today's socio-technical systems requires a complex systems approach which allows us to explicitly capture the multi-disciplinary nature of infrastructures, industrial systems and services.

While many elements of our work have been published in PhD theses, book chapters, journals and proceedings of international conferences, until now there was no single resource which described our common view, presented the methodology which emerged from our work, and collected the lessons learned modelling as a team. In this volume we have collected the background, theory and practical steps and brought them together with a number of inspiring case studies that show how this approach works in practice and how such models can support decision-making.

All authors who contributed to this book played an important role in developing, crystallising and testing the methodology. So first of all, we would like to thank them for their valuable contributions and for sharing their experience. Furthermore, we would like to thank the Next Generation Infrastructures Foundation and the Delft University of Technology for the financial support of our research projects and for making it possible to write this book. As editors we are more than grateful for the support we received from Deborah Sherwood in proofreading and language editing the texts.

Finally, one small comment about the use of the word *agent*. We consider an agent to be a *representation* of a decision-making entity in the real world, be it an individual or an organisation, and it is a stylised part of the model. As such, throughout this book we have referred to an agent as 'it' rather than with a personal pronoun.

We hope students, researchers and business professionals using this book will appreciate the theoretical base and practical guidelines for developing agent-based models and that it will help successfully cope with the complexities of the design and management of socio-technical systems.

Delft, the Netherlands

Koen H. van Dam
Igor Nikolic
Zofia Lukszo

Contents

1 Introduction . . . 1
G.P.J. Dijkema, Z. Lukszo, and M.P.C. Weijnen
1.1 Why This Book? . . . 1
1.2 Infrastructures as Complex Adaptive Socio-technical Systems . . . 2
1.3 Better Decision-Making Needed . . . 4
1.4 Agent-Based Modelling for Decision Support . . . 5
1.5 A Book in Two Parts . . . 7
References . . . 7

Part I Theory and Practicc

2 Theory . . . 11
I. Nikolic and J. Kasmire
2.1 Introduction . . . 11
2.1.1 Focus . . . 12
2.1.2 Structure of the Chapter . . . 12
2.1.3 Example: Westland Greenhouse Cluster . . . 13
2.2 Systems . . . 14
2.2.1 History of Systems Thinking . . . 14
2.2.2 Systems . . . 16
2.2.3 World Views . . . 19
2.2.4 Observer-Dependence . . . 21
2.2.5 System Boundaries . . . 24
2.2.6 System Nestedness . . . 25
2.3 Adaptive . . . 26
2.3.1 Adaptation Versus Evolution . . . 27
2.3.2 Evolution—More than just Biology . . . 28
2.3.3 Adaptation in Its Many Forms . . . 30
2.3.4 Direction of Adaptation . . . 31
2.3.5 Coupled Fitness Landscape . . . 32
2.3.6 Intractability . . . 34

2.4 Complexity . . . 36
2.4.1 Simple . . . 36
2.4.2 Complicated . . . 39
2.4.3 Complex . . . 41
2.5 Complex Adaptive Systems . . . 44
2.5.1 Chaos and Randomness . . . 45
2.5.2 Emergence, Self-organisation and Patterns . . . 48
2.6 Modelling Complex Adaptive Systems . . . 51
2.6.1 What Does a Model of a Complex Adaptive System Need? . . . 51
2.6.2 Agent-Based Modelling . . . 54
2.6.3 What It Is and Is not . . . 55
2.7 Anatomy of an Agent-Based Model . . . 57
2.7.1 Agent . . . 57
2.7.2 Environment . . . 61
2.7.3 Time . . . 65
References . . . 68

3 **Practice** . . . 73
I. Nikolic, K.H. van Dam, and J. Kasmire
3.1 Introduction . . . 73
3.2 Step 1: Problem Formulation and Actor Identification . . . 74
3.2.1 Step 1 Example . . . 75
3.3 Step 2: System Identification and Decomposition . . . 77
3.3.1 Inventory . . . 77
3.3.2 Structuring . . . 78
3.3.3 Step 2 Example . . . 81
3.4 Step 3: Concept Formalisation . . . 82
3.4.1 Software Data Structures . . . 83
3.4.2 Ontology . . . 84
3.4.3 Step 3 Example . . . 85
3.5 Step 4: Model Formalisation . . . 88
3.5.1 Developing a Model Narrative . . . 88
3.5.2 Pseudo-code . . . 89
3.5.3 Step 4 Example . . . 92
3.6 Step 5: Software Implementation . . . 93
3.6.1 Modelling Environment . . . 94
3.6.2 Programming Practices . . . 95
3.6.3 Step 5 Example . . . 98
3.7 Step 6: Model Verification . . . 98
3.7.1 Recording and Tracking Agent Behaviour . . . 100
3.7.2 Single-Agent Testing . . . 101
3.7.3 Interaction Testing in a Minimal Model . . . 102
3.7.4 Multi-agent Testing . . . 103
3.7.5 Step 6 Example . . . 104

3.8 Step 7: Experimentation . . . 105
3.8.1 Experiment Design Aspects . . . 105
3.8.2 Experiment Setup . . . 108
3.8.3 Experiment Execution . . . 112
3.8.4 Step 7 Example . . . 114
3.9 Step 8: Data Analysis . . . 116
3.9.1 Data Exploration . . . 116
3.9.2 Pattern Visualisation and Identification . . . 120
3.9.3 Pattern Interpretation and Explanation . . . 123
3.9.4 Experiment Iteration . . . 123
3.9.5 Step 8 Example . . . 124
3.10 Step 9: Model Validation . . . 126
3.10.1 Historic Replay . . . 127
3.10.2 Expert Validation . . . 128
3.10.3 Validation by Literature Comparison . . . 129
3.10.4 Validation by Model Replication . . . 129
3.10.5 Step 9 Example . . . 130
3.11 Step 10: Model Use . . . 130
3.11.1 Outcome Presentation . . . 131
3.11.2 Raising New Questions . . . 131
3.11.3 Long Term Stakeholder Engagement . . . 132
3.11.4 Agent-Based Models and Stakeholders . . . 132
3.11.5 Computer Models and Mental Models . . . 133
3.11.6 Step 10 Example . . . 134
3.12 Chapter Conclusions . . . 135
References . . . 136

Part II Case Studies

4 Introduction to the Case Studies . . . 141
K.H. van Dam, I. Nikolic, and Z. Lukszo
4.1 Case Studies . . . 141
4.1.1 Operational Models . . . 142
4.1.2 Tactical Models . . . 142
4.1.3 Strategic Models . . . 142
4.1.4 Setup of Case Study Chapters . . . 143
4.1.5 State-of-the-Art Modelling . . . 144
4.2 An Ontology for Socio-technical Systems . . . 144
4.2.1 Aims . . . 144
4.2.2 Development . . . 145
4.2.3 Key Concepts . . . 146
4.2.4 Usage . . . 147
References . . . 149

5 Agent-Based Models of Supply Chains . . . 151
B. Behdani, K.H. van Dam, and Z. Lukszo
5.1 Introduction . . . 151
5.1.1 Supply Chains . . . 152
5.1.2 Abnormal Situation Management . . . 152
5.1.3 Supply Chain Modelling . . . 153
5.1.4 Chapter Organisation . . . 154
5.2 Oil Refinery Supply Chain . . . 154
5.2.1 Step 1: Problem Formulation and Actor Identification . . . 154
5.2.2 Step 2: System Identification and Decomposition . . . 155
5.3 Multi-plant Enterprise Supply Chain . . . 159
5.3.1 Step 1: Problem Formulation and Actor Identification . . . 159
5.3.2 Step 2: System Identification and Decomposition . . . 160
5.4 Step 3: Concept Formalisation . . . 160
5.5 Step 4: Model Formalisation . . . 164
5.5.1 Procurement . . . 164
5.5.2 Order Assignment . . . 165
5.6 Step 5: Software Implementation . . . 167
5.7 Step 6: Model Verification . . . 167
5.8 Step 7: Experimentation . . . 169
5.8.1 Experimental Setup for the Oil Refinery Supply Chain . . . 169
5.8.2 Experimental Setup for the Multi-plant Enterprise . . . 172
5.9 Step 8: Data Analysis . . . 173
5.9.1 Delay in Shipment in the Oil Refinery Supply Chain . . . 173
5.9.2 Normal and Abnormal Behaviour Analysis for the Multi-plant Enterprise . . . 174
5.10 Step 9: Model Validation . . . 177
5.11 Step 10: Model Use . . . 177
5.12 Conclusions . . . 178
References . . . 179

6 An Agent-Based Model of Consumer Lighting . . . 181
E.J.L. Chappin and M.R. Afman
6.1 Introduction . . . 181
6.2 Step 1: Problem Formulation and Actor Identification . . . 183
6.3 Step 2: System Identification and Decomposition . . . 183
6.3.1 Inventory . . . 183
6.3.2 Structuring . . . 187
6.4 Step 3: Concept Formalisation . . . 188
6.5 Step 4: Model Formalisation . . . 190
6.5.1 Social Structure . . . 190
6.5.2 Model Narrative and Pseudo Code . . . 191
6.6 Step 5: Software Implementation . . . 193
6.7 Step 6: Model Verification . . . 194
6.8 Step 7: Experimentation . . . 194

6.9 Step 8: Data Analysis 195
6.10 Step 9: Model Validation 197
6.11 Step 10: Model Use 198
6.12 Conclusions 198
References 199

7 An Agent-Based Model of CO_2 Policies and Electricity Generation . 201
E.J.L. Chappin and G.P.J. Dijkema
7.1 Introduction 201
7.2 Step 1: Problem Formulation and Actor Identification 202
7.3 Step 2: System Identification and Decomposition 204
7.4 Step 3: Concept Formalisation 206
7.5 Step 4: Model Formalisation 206
7.5.1 Model Narrative 207
7.5.2 Investment 208
7.5.3 Bidding on Markets 208
7.6 Step 5: Software Implementation 210
7.7 Step 6: Model Verification 211
7.7.1 Preliminary Simulations 211
7.7.2 Extreme-Condition Tests and Discussion 212
7.7.3 Agent Behaviour Tests 212
7.7.4 Repetitions 213
7.8 Step 7: Experimentation 213
7.9 Step 8: Data Analysis 214
7.10 Step 9: Model Validation 215
7.11 Step 10: Model Use 215
7.11.1 Emergent Insights from Iterations and Discussions 216
7.11.2 Serious Game 216
7.12 Conclusions 217
References 218

8 An Agent-Based Model of a Mobile Phone Production, Consumption and Recycling Network 221
L.A. Bollinger, C.B. Davis, and I. Nikolic
8.1 Introduction 221
8.2 Step 1: Problem Formulation and Actor Identification 223
8.3 Step 2: System Identification and Decomposition 223
8.4 Step 3: Concept Formalisation 226
8.5 Step 4: Model Formalisation 228
8.5.1 Create Contracts 228
8.5.2 Purchase 229
8.5.3 Process 229
8.5.4 Invest 231
8.6 Step 5: Software Implementation 231
8.7 Step 6: Model Verification 231
8.8 Step 7: Experimentation 233

8.9 Step 8: Data Analysis 235
8.9.1 Default Case 235
8.9.2 Sweep 1 236
8.9.3 Sweep 2 237
8.9.4 Sweep 3 239
8.10 Step 9: Model Validation 239
8.11 Step 10: Model Use 240
8.12 Conclusions 241
References 242

9 Next Steps in Modelling Socio-technical Systems: Towards Collaborative Modelling 245
A. Chmieliauskas, C.B. Davis, and L.A. Bollinger
9.1 Complications of Modelling Socio-technical Systems 245
9.1.1 Agent-Based Model as Software 246
9.2 Applying Semantic Web Technologies and Methods for Modelling 247
9.2.1 Semantic Web Technologies 247
9.2.2 Using SPARQL to Extract Structured Data 249
9.2.3 Using SPARQL Within Simulations 250
9.2.4 Implementations by Other Researchers 253
9.3 Case Study: Mobile Phone Recycling Networks 253
9.4 Future Work: Enabling Collaboration Between Modellers 257
9.4.1 Limitations of Using a Single Ontology for Modelling . . . 257
9.4.2 Enabling Multiple System Views 258
9.4.3 Integrating Multiple Ontologies Together for Simulations . 259
9.4.4 Simple Example of Integrating Multiple Ontologies 259
9.5 Conclusion 262
References 262

Index 265

Contributors

Maarten R. Afman

Maarten Afman studied Systems Engineering, Policy Analysis and Management at Delft University of Technology where he graduated developing an agent-based modelling approach of the consumer lighting sector. Maarten has a background in ICT (ICT services consultancy; programming) and Applied Physics (Delft University of Technology). Currently, Maarten works as an energy policy consultant for CE Delft, an environmental policy think tank based in the Netherlands. At CE he works on energy efficiency in the industrial and ICT sectors, economic analyses of energy options and assessments of environmental effects using methods such as life cycle analysis (LCA) and CO_2-footprinting.

email: m.afman@gmail.com
www: http://www.ce.nl/index.php?go=home.showPages&pagenr=630

Behzad Behdani

Behzad Behdani is currently working as a PhD researcher in the Faculty of Technology, Policy and Management at Delft University of Technology, the Netherlands. He received his MSc in Energy System Engineering from Sharif University of Technology, Iran in 2006. Behzad was working as a researcher in several research institutes and consultancy companies before joining Delft University of Technology in 2008. The main focus of his PhD research is on managing disruptions and abnormalities in supply chains and industrial networks. His other research interests include logistics and supply chain management, supply chain modelling and simulation, agent-based modelling and complex systems theory.

email: b.behdani@tudelft.nl

L. Andrew Bollinger

Andrew Bollinger is a PhD candidate at Delft University of Technology in the Netherlands. He holds an MSc degree in Industrial Ecology (2010) and a BSc degree in Engineering/Economics from Dartmouth College in the United States (2002). From 2003 to 2005, Andrew worked with EPEA GmbH, an environmental design consultancy in Hamburg, Germany. After teaching at a technical university in Hangzhou, China in 2005–2006, Andrew undertook an MSc degree in Industrial Ecology at Delft University of Technology and Leiden University in the Netherlands. In 2009–2010, he completed his MSc project using a combination of agent-based modelling and system dynamics to investigate the global flows of metals in mobile phones. Since 2010, Andrew has been a PhD candidate within the Faculty of Technology, Policy and Management at Delft University of Technology. His current research employs agent-based modelling to investigate the adaptation of electricity infrastructures to climate change.

email: L.A.Bollinger@tudelft.nl
www: http://wiki.tudelft.nl/bin/view/Main/AndrewBollinger

Émile J.L. Chappin

Émile Chappin obtained his BSc and MSc degrees in Systems Engineering, Policy Analysis and Management at the Faculty of Technology, Policy and Management of Delft University of Technology. In 2007, he worked part-time for the Centre for Environmental Sciences of Leiden University on a life cycle assessment of bio-energy. In 2011 Émile obtained his PhD from Delft University of Technology with a thesis titled “Simulating Energy Transitions” (available online http://chappin.com/thesis), with a focus on agent-based modelling and serious gaming of long-term, structural changes in energy infrastructures. Émile is currently doing a postdoc at Delft University of Technology, among other things on the adaptation of energy infrastructures to the potential effects of climate change. In addition, he is a guest researcher at the Wuppertal Institute for Climate, Environment and Energy in Germany.

email: e.j.l.chappin@tudelft.nl
www: http://chappin.com/emile

Alfredas Chmieliauskas

Alfredas Chmieliauskas is a PhD candidate at the Energy and Industry group. He received his MSc in Economics and Finance from Stockholm School of Economics, Sweden in 2004. After spending a few years doing economic research for the financial industry he decided to join the Energy and Industry group at Delft University of Technology to explore the joint evolution of human behaviour and large-scale infrastructure. His research spans multiple domains, including knowledge management, crowd sourcing, modelling and simulation, renewable and fossil based energy sources, infrastructures and markets.

email: a.chmieliauskas@tudelft.nl
www: http://enipedia.org

Chris B. Davis

Chris Davis is currently a PhD Candidate at the Energy and Industry group, Faculty of Technology, Policy and Management, at the Delft University of Technology. In 2001, he graduated with a Bachelor of Engineering in Electrical Engineering and Computer Science from Vanderbilt University. In 2007, he received MSc in Industrial Ecology from Leiden University with a thesis combining life cycle assessment within agent-based modelling. His current work involves tackling issues of sustainability through a combination of tools such as the semantic web, agent-based models, and collaborative software such as wikis.

email: c.b.davis@tudelft.nl
www: http://wiki.tudelft.nl/bin/view/Main/ChrisDavis
http://enipedia.tudelft.nl

Gerard P.J. Dijkema

Gerard Dijkema is Associate Professor in the Energy and Industry group, Faculty of Technology, Policy and Management, Delft University of Technology. His specialisation is system innovation for sustainability. With a group of talented researchers, he focuses on the understanding, development and transition of energy infrastructure and industrial networks. Core theme of his work is modeling the evolution of socio-technical systems to analyse and increase our understanding of the interplay of technology, economics, policy and regulation that affect performance, structure and content of these systems and provide decision support. A leader of various projects in the Next Generation Infrastructure programme, he co-developed the EDGaR programme that addresses gas technology, systems and markets and the INCAH project, which focuses on infrastructure adaptation to climate change. He graduated as a Chemical Engineer (hons.) from Twente University of Technology and holds a PhD ("Process System Innovation by Design—Towards a Sustainable Petrochemical Industry") from Delft University of Technology. He is editor of the Journal of Industrial Ecology and a Member of the General Council of the Delfland Water Authority. Working with Shell he obtained industrial experience.

email: g.p.j.dijkema@tudelft.nl

Julia Kasmire

Julia Kasmire is currently a PhD Candidate at the Energy and Industry group, Faculty of Technology, Policy and Management, at the Delft University of Technology. After obtaining a BSc degree in Linguistics from the University of California, Santa Cruz, in 2004 Julia went on to get an MSc in the Evolution of Language and Cognition from the University of Edinburgh in Scotland. Her thesis there focused on the effects of modelling parameters on simulations of language evolution. Her current work uses the Westland greenhouse horticultural sector in the Netherlands as a case study to examine the evolution of technologies, including innovation, diffusion and competition, through agent-based models. She also investigates how proposed policies and regulations are likely to affect the course of technological development and implementation and whether the interventions are likely to function as desired.

email: j.kasmire@tudelft.nl

Zofia Lukszo

Zofia Lukszo is an Associate Professor at Delft University of Technology, the Netherlands. She acquired her PhD from Eindhoven University of Technology, the Netherlands. Her research focuses on operations management, mainly in the process industry and infrastructure sectors. She has extensive experience in the application of mathematical modelling, optimisation and quality control in decision making at the operation and control level in the process industry. She is a leader of the programme Intelligent Infrastructures within the international research programme of the Next Generation Infrastructures (http://www.nextgenerationinfrastructures.eu/).

email: z.lukszo@tudelft.nl
www: http://tbm.tudelft.nl/index.php?id=31984

Igor Nikolic

Igor is an Assistant Professor at the Energy and Industry group, Faculty of Technology, Policy and Management faculty, Delft University of Technology. In his research he specialises in applying complex adaptive systems theory, agent-based modelling and evolutionary theory to model industry and infrastructure network evolution. He takes a heavy hint from evolutionary biology and ecosystem behaviour in his understanding of industrial ecology and socio-technical system evolution. He is an active networker and promoter of open source and social software that enables collaborative, multidisciplinary research work. Igor graduated for his MSc with honours as a chemical- and bio-process engineer from Delft University of Technology. In his MSc thesis he presented an agent-based model of gene flow from GMO crops to surrounding plant populations. After his graduation, he spent several years as an environmental researcher and consultant at University of Leiden, Institute for Environmental Science (CML), where he worked on life cycle analysis (LCA), material/substance flow analysis (MFA/SFA) and has specialised in industrial ecology. He has received his PhD thesis in November 2009. His thesis work focused on the design of a co-evolutionary method for constructing agent-based models of the evolution of Large Scale Socio-Technical systems.

email: i.nikolic@tudelft.nl
www: http://www.igornikolic.com

Koen H. van Dam

Koen van Dam holds an MSc in Artificial Intelligence from the Vrije Universiteit in Amsterdam and a PhD from Delft University of Technology for his thesis entitled "Capturing socio-technical systems with agent-based modelling". Afterwards he worked as post-doc in the Energy and Industry group at the faculty of Technology, Policy and Management of the Delft University of Technology on modelling socio-technical infrastructure systems, including energy markets and supply chains. He was visiting researcher in the Urban Energy Systems project at Imperial College as well as at the Department of Chemical and Biomolecular Engineering of the National University of Singapore. He was also the president of Eurodoc, the European council for doctoral candidates and young researchers, and board member of other organisations representing early stage researchers. Currently he is a research fellow at Imperial College London working on the Digital City Exchange project. His main research interests include cities and urban planning, infrastructure systems, digital services, agent-based modelling, consumer acceptance of technology and ontology design.

email: k.van-dam@imperial.ac.uk
www: http://www.koenvandam.com

Margot P.C. Weijnen

Margot Weijnen is the founding and scientific director of the Next Generation Infrastructures Foundation, established in 2001. The Next Generation Infrastructures knowledge programme focuses on issues at the interface of public policy and technological change in the public utilities and network industries. It is executed by an international consortium of universities, governmental organisations and private partners with financial support by the Dutch government. Margot Weijnen is a full professor of process and energy systems engineering at TU Delft, in the Department of Technology, Policy and Management, since 1995. From 1990 to 1995 she headed the interfaculty Delft University Clean Technology institute. Before 1990, she worked in R&D, chemicals manufacturing operations and process design engineering for Royal Dutch Shell. She obtained her MSc and PhD degrees at TU Delft in the field of chemical process engineering. Margot Weijnen served and serves in numerous advisory functions for the Dutch government, infrastructure network operators and the European Commission, especially in the areas of energy, technology and innovation policy. Among other affiliations, she is a member of the Executive Board of the Netherlands Technology Foundation (STW) and a member of the Supervisory Board of AkzoNobel Nederland BV.

email: m.p.c.weijnen@tudelft.nl
www: http://tbm.tudelft.nl/index.php?id=32066

List of Figures

Fig. 2.1 Observer-dependence . . . 21
Fig. 2.2 Deformation of a fitness landscape . . . 33
Fig. 2.3 An intractibility path . . . 35
Fig. 2.4 Power law observed in the edit frequency per user on wiki.tudelft.nl . . . 43
Fig. 2.5 The Mandelbrot set fractal . . . 46
Fig. 2.6 Structure of an agent-based model . . . 58

Fig. 3.1 Greenhouse example schematic model layout . . . 82
Fig. 3.2 Example ontology of Currencies as a Unit . . . 86
Fig. 3.3 Segment of the example ontology for Growers . . . 87
Fig. 3.4 Latin Hypercube Sampling vs Random Sampling . . . 110
Fig. 3.5 Different types of visualisations . . . 121
Fig. 3.6 The total count of companies with each type of technology in the greenhouse water technology category . . . 126

Fig. 4.1 The relationship between different classes of Nodes . . . 147
Fig. 4.2 OperationalConfigurations for the class Technology . . . 148
Fig. 4.3 Physical edges in the ontology . . . 149

Fig. 5.1 Schematic depiction of an oil refinery supply chain . . . 157
Fig. 5.2 Main actors and their behaviours/interactions in a lube oil multi-plant enterprise . . . 161
Fig. 5.3 An Ownership edge connecting a Social Node and a Physical Node . . . 162
Fig. 5.4 Order ontology . . . 163
Fig. 5.5 Model behaviour (inventory level for production plant 1) under different extreme condition tests . . . 168
Fig. 5.6 Results from the oil refinery supply chain case study . . . 174
Fig. 5.7 The effect of the procurement policy on total tardiness and number of late orders . . . 175

Fig. 5.8 The effect of the procurement policy on total tardiness and late orders during an abnormal situation . . . 176
Fig. 5.9 The effect of the procurement policy on average inventory level during an abnormal situation . . . 177

Fig. 6.1 Overview of the main technological components in consumer lighting . . . 185
Fig. 6.2 Structured overview of components and interactions in the consumer lighting system . . . 187
Fig. 6.3 Extension of the shared ontology for socio-technical systems with case-specific concepts . . . 189
Fig. 6.4 Conceptual view of a household's data structure concerning lamps . . . 190
Fig. 6.5 Scale free network of households . . . 191
Fig. 6.6 Flow chart representing the sequence of actions during the household's step . . . 192
Fig. 6.7 The average electricity intensity of the modelled households . . . 195
Fig. 6.8 Total number of lamps in the simulation . . . 196
Fig. 6.9 Total number of lamps in the simulation under different settings . 197

Fig. 7.1 Socio-technical systems perspective on power generation and emissions policy . . . 205
Fig. 7.2 Formalisation of agent decision-making in three layers . . . 207
Fig. 7.3 Iterative clearing of power and CO_2 markets . . . 209
Fig. 7.4 Software model development and running and analysing simulations . . . 211
Fig. 7.5 Electricity prices, CO_2 prices and emission levels . . . 214
Fig. 7.6 Average generation portfolio evolution for the three scenarios . . . 214

Fig. 8.1 Elements and relationships associated with agent states and behaviours . . . 225
Fig. 8.2 Elements and relationships associated with mobile phones . . . 226
Fig. 8.3 State of phones in the system over time for the default case . . . 236
Fig. 8.4 Indicator values produced by variation of the parameter in isolation . . . 237
Fig. 8.5 Distribution of CLI values over all simulation runs . . . 238
Fig. 8.6 CLI value vs. mass of gold disposed for all simulation runs . . . 238
Fig. 8.7 Assets of the refurbisher agent over 100 runs at a single set of parameter values . . . 240

Fig. 9.1 Behaviour, query and data . . . 246
Fig. 9.2 Depiction of the ontology used for many of the models described in this book . . . 247
Fig. 9.3 The Semantic Web technology stack . . . 248
Fig. 9.4 A triple—the atom of the Semantic Web . . . 248

Fig. 9.5 RDF graph example 248
Fig. 9.6 The traditional and proposed approach to agent-based modelling with ontologies 250
Fig. 9.7 Output from a SPARQL query used for debugging a model run . . 255
Fig. 9.8 Useful model decomposition following the MVC analogy 259
Fig. 9.9 Shared data graph 260
Fig. 9.10 Resulting data graph 261

List of Tables

Table 5.1 Agents and Physical Nodes in the refinery supply chain case . . . 158
Table 5.2 Agents and Physical Nodes in the multi-plant enterprise case . . . 162
Table 5.3 Outcome of the Nelder-Mead simplex optimisation 175
Table 5.4 Performance data for the enterprise in a normal and an abnormal situation (Plant 1 shutdown) 176

Table 7.1 Agents, technologies in the emission policies model 205

Table 8.1 List of agent types in the model and the technologies they possess 224
Table 8.2 Properties and possible states of mobile phones 227
Table 8.3 List of parameters 234

Chapter 1
Introduction

G.P.J. Dijkema, Z. Lukszo, and M.P.C. Weijnen

Abstract This chapter provides an introduction to agent-based modelling of socio-technical systems. The challenges faced by actors in various infrastructure sectors are illustrated with a range of problems in the way they are operating today and how they may evolve over time. These problems are not easy to solve because such systems have many components and levels, involving different parties who all primarily pursue their own local objectives in a dynamic environment and regulatory regime. Adopting a socio-technical system view implies that a social network of actors and a physical network of technical artefacts together form a complex adaptive system: a multi-actor network determines the development, operation and management of the technical network, which in turn affects the behaviour of the actors. Simulating the system with the agent-based modelling approach and conducting "in silico" experiments helps in understanding this complexity and provides decision support for steering these complex socio-technical systems.

1.1 Why This Book?

This book provides a comprehensive, hands-on introduction to agent-based modelling of socio-technical systems. Adopting a socio-technical systems view implies that we model parts of the man-made world (Allenby 2006; Crutzen and Stoermer 2000) as systems composed of two deeply interconnected subsystems: a social network of actors and a physical network of technical artefacts. Together, these intertwined systems form a complex adaptive system: a multi-actor network determines the development, operation and management of the technical network, which in turn affects the behaviour of the actors. The interactions within and between technical systems are defined by causal relationships which are governed by laws of nature, while the actors in the social system develop intentional relationships to accomplish their individual goals. At multiple hierarchical levels the technical network is shaped by the social network and vice-versa, with feedback loops running across

Z. Lukszo (✉)
http://tbm.tudelft.nl/index.php?id=31984
e-mail: z.lukszo@tudelft.nl

K.H. van Dam et al. (eds.), *Agent-Based Modelling of Socio-Technical Systems*,
Agent-Based Social Systems 9,
DOI 10.1007/978-94-007-4933-7_1, © Springer Science+Business Media Dordrecht 2013

multiple levels and time scales. All of this together forms a self-organising, hierarchical, open system[1] with a multi-actor, multi-level and multi-objective character (Kay 2002; Holland 1992).

This book is essentially about improving decision-making, in both the public and private arenas. This is important because the operation and the evolution of socio-technical systems are constrained by as well as driven by decisions. Improvement involves acknowledging the real-world complexity of such a system and experimenting with alternative strategies and institutions to discover their potential effects on the behaviour of the overall system before implementing them.

The primary intended audience of this book includes the actors that operate (large parts of) the socio-technical system, the actors that make large infrastructure investments, and the actors that draft new legislation and regulations. All of them are in need of a new way to see how their decisions, and all changes affecting user preferences, play out with regard to accomplishing public interests and private goals.

1.2 Infrastructures as Complex Adaptive Socio-technical Systems

Let us begin our description of socio-technical systems by discussing a typical example: the electricity infrastructure. The physical electricity infrastructure provides electric power; it contains millions of technological artefacts, such as wind and solar energy-harvesting facilities, thermal power plants, transmission lines, transformer stations and distribution networks. It is an open system, meaning that fuels and renewable energy sources are taken in, allowing electricity flows to be generated and transported, while heat, emissions and other waste products are dissipated into the natural environment. In terms of a social system, the electricity infrastructure involves many actors, ranging from power generators, transmission and distribution network operators, power suppliers, industrial users and consumers, to electricity market operators, traders and brokers, policy makers and regulatory authorities. The operational decisions and investment strategies of these actors are governed by institutions at the local level, by national legislation and regulation, and by supranational directives and agreements, such as the European directives on electricity sector reform and the Kyoto Agreement on the reduction of global carbon dioxide emissions. Some of the actors are multinational companies, but even the actors operating on a local or national scale are affected by supranational institutions and by decisions taken by other actors abroad.

The social infrastructure network can be decomposed into public and private actors, populating different segments of the value chain, at different hierarchical levels, and with different objectives and resources to reach their short-term and long-term goals. Depending on the research question at hand, we may want to formulate more detailed distinctions between actors, for example through grouping consumers by income, environmental awareness or other lifestyle characteristics. We could also

[1] A so-called SOHO system.

conduct an in-depth exploration of their decision-making rules as set by, for example, market forces, the legislative and regulatory framework or the broader institutional context (de Bruijn and Herder 2009; Dijkema and Basson 2009). Last but not least, we must recognise that the system is constantly in flux: the electricity infrastructure evolved over many decades, constantly being re-shaped by the decisions of many actors in response to changes in economic conditions, user demands, societal priorities and institutional frameworks.

The electricity infrastructure arrangement that dominated in the 20th century materialised because fossil resources, technology and economics favoured a vertically-integrated monopoly setup in which electric power is provided by relatively few large producers to many medium and small consumers. Enabled by technological advancement, driven by concerns about climate change, dependency on foreign oil and gas and increasing energy cost, and accelerated by the liberalisation and deregulation of the electricity market, in just a decade distributed generation has taken off, utilising a mix of both fossil and renewable resources. Combined with the ongoing reregulation of the electricity sector, this has created both short-term and long-term uncertainty and has forced network operators, power companies, and industrial producers and consumers to carve out and redefine their niches in the power sector ecosystem. Today consumers produce their own power, institutional investors and green energy collectives build wind parks, and greenhouse operators and a host of other actors leverage small- to medium-scale generation technologies such as wind turbines, photovoltaics and (micro) combined heat and power.

It is thus apparent that within a changing institutional framework, the system is self-organising in a process that co-evolves with the web of institutions and regulations. This process of self-organisation occurs at multiple levels: the system is composed of national, regional and local subsystems including generation, transmission, distribution and loads. With an increasing share of generation capacity from wind farms, and because an increasing number of power consumers can switch roles to become power producers (so-called "prosumers"), the amount and the quality of power produced in such a decentralised system can vary enormously. The number and types of links in the electricity infrastructure system are continuously increasing, each and every link potentially affecting the real-time integrity of the grid and its long-term evolution. The shift from centralised to decentralised power generation, the adoption of small-scale renewable power generation units, the introduction of electric vehicles and the implementation of smart grids are some of the prominent developments that dramatically affect the structure and behaviour of the electricity infrastructure, in both the technical and social dimensions.

The multi-actor, multi-level and multi-objective character of infrastructure systems can also be illustrated by the road infrastructure. This system consists of different networks (motorways, urban and rural roads), and within each of these networks three distinct levels can be distinguished: the network level, the road-segment level and the point (or cross-section) level. At the national and supra-national level, policies apply to the motorway system which guide its development and use for years at a time. Inherent to the capital intensity of the physical system, decisions to change these policies take shape over years and even decades. At the other end of the spectrum, many of the decisions taken by individual drivers, such as their choice of speed or lane, unfold at the point level and on a time scale of seconds to minutes.

At the national and supra-national level, policy making is typically a process that serves to arrive at balanced multi-criteria decisions, with safety, network performance, environmental protection and travel time reliability being the most common criteria. In addition, interdependencies between road infrastructure planning and other policy areas, such as spatial planning and economic development, must be taken into account. Although some objectives are more or less in harmony (e.g. network performance improves if there are fewer accidents), others are conflicting (e.g. imposing a very low speed limit would virtually eliminate fatal accidents but would at the same time make the network performance unacceptable). There is a tendency to solve this type of problem by reducing complexity and focusing on a simplified single-level cost/benefit analysis in which each of the criteria is represented monetarily (Lukszo et al. 2006). Yet it is possible to address these issues at a more complex network level. Snelder (2010), for example, concluded that local and regional traffic should be separated from long-distance traffic on local highway segments, to improve both highway and local network performance. The logical next step would be to move from separate multi-objective, multi-level or multi-actor analyses to a more comprehensive analysis that combines multi-level and multi-actor and multi-criteria aspects, thereby supporting actors in handling the real complexity of infrastructure planning and helping them to take better decisions.

1.3 Better Decision-Making Needed

From the examples of the electricity and transport infrastructures it is clear that both the real-time performance of these systems and their long-term evolution are shaped by a myriad of actors, with none of them being in a position to control the whole system. Moreover, the ensuing interactions and interdependencies entail new, unknown and possibly unacceptable risks, which add an extra dimension to the system's complexity which needs to be accounted for during the decision making process. The stakes are huge. For example, whenever the regulatory regime is changed, some actors reap the benefits while others bear the consequences. Each and every decision may represent the proverbial butterfly and set great things in motion. But what are these things, and in which direction do they play out?

Rather than focusing on specific system components or subsystems, our efforts should be aimed at understanding and steering the structure and behaviour of the system as a whole. The system representation should not be confined to technological aspects but should also address the social dimension, seeking to capture the behaviour of different actors in decision making, in competing or in co-operating and negotiating. The concept of agent-based systems, composed of multiple interacting actors and physical elements, is a promising modelling approach that can simulate how system behaviour emerges from the behaviour of actors at the bottom level.

With the assumption that we can indeed capture the behaviour of real actors, agent-based simulations allow us to observe how the technical and social subsystems of an infrastructure co-evolve, and which overall system behaviour might emerge from their ongoing interactions, at multiple system levels and time scales. The importance of these simulations is that they can also help to answer the question of what a "better" system might be. Systems can be improved in a variety of ways, the appreciation of which is strongly dependent on the perspective of the decision maker. Generally speaking, in day-to-day operation, improvement is concerned with system performance, including efficiency and effectiveness. Over a longer time frame, systems also need to be robust and resilient. They need to be responsive, flexible and adaptable. Negative external effects manifest themselves on a variety of spatial and time scales. Socio-economic effects ripple through intensively interconnected social systems, financial markets, international supply chains, geopolitics, etc. Hence, we must accept the fact that there are no easy, isolated solutions or quick fixes; there is much to be learned. Agent-based models contribute to the process of learning and thereby to finding more complete answers.

1.4 Agent-Based Modelling for Decision Support

In agent-based modelling, an agent is the software representation of some entity that completes an action or takes a decision, by which it effectively interacts with its environment. The agent may represent an actor in a social network (e.g. a human being, an operator deciding to shut down a pumping station, or a consumer who buys tomatoes). Beyond this level of granularity, an agent may represent an organisation, for example an electric utility company deciding to build a new nuclear plant or a government deciding on new policy to ensure drug safety. The agent paradigm aligns with the concept of systems composed of multiple interacting social entities and technical subsystems. As such, it is the premier candidate with which to model socio-technical systems and explore the dynamics and structural change ensuing from the interactions within and between the social and the technological networks.

Agent-based concepts generally are the best means by which to model complex systems, as long as the following conditions are satisfied (van Dam 2009):

- The problem has a distributed character; each actor is to some extent autonomous.
- The subsystems (agents) operate in a highly dynamic environment.
- Subsystem interaction is characterised by flexibility. It can result from a reactive or pro-active attitude, from a propensity to co-operate or to compete, or it can be the result of social interaction (including, for example, trust or empathy).

As mentioned above, many large systems created by humans can be considered to be socio-technical systems. When we suitably delineate such systems and decide on a system boundary, we can take the next step with agent-based models (Nikolic 2009), namely conducting experiments "in-silico". By simulating their operation and evolution we can increase our understanding of socio-technical systems, adjust model parameters to reproduce historic developments, elucidate the drivers, scope and limits of a system's evolution, and visualise possible evolutionary pathways or the outcomes of operational strategies. Through the use of computers, running multiple simulations with agent-based models allows us to explore the relevant "decision space", thus completing rich ex-ante analyses of the possible outcomes and elucidating system characteristics before creating the real system. This can be particularly advantageous when systems are developed over the course of years or decades, or when the risks associated with incorrect operational decisions are large.

It should be stressed that even when system content and structure are "fixed", socio-technical systems are not static. Rather, they exhibit complex dynamic behaviour. Using agent-based modelling, we can develop simulations to study the operation of such systems and the behaviour that results from the real or simulated interactions between multiple users of an industrial network or its products, or of energy or transport infrastructure. The key is that by decomposing the system into agents that interact, achieve input/output conversions or take decisions, we set the stage for being able to generate system behaviour by merely defining the systems' constituents and their interactions. An agent-based model attempts to produce realistic—though not necessarily reliable(!)—results for dynamic, evolving systems that are subject to control or intervention by more than a single person or organisation. An agent-based model allows one to simulate the operation, dynamics, evolution and growth of a system.

Essentially, in the agent paradigm, everything is a "thing that interacts with other things" (Rohilla Shalizi 2006). All things together make up the system. In an agent-based model, we model the "things" and their interactions. These can be either technical objects, such as production installations, electric vehicles, gas pipelines or control systems, or social entities including individuals, organisations or parts thereof. For any agent-based simulation model, we thus need to capture the relevant behaviour and applicable interactions. Once that has been accomplished, we can encode them into a simulation model and determine or generate the starting condition. Upon pressing "run" the agents will begin to interact and form linkages, and we can watch the system evolve. As a matter of course, apart from the behaviour and interaction, we can also track all kinds of characteristics of the agents and visualise them per agent or system component or in terms of the evolved system.

Whether or not a system with favourable characteristics emerges is subject to a myriad of parameters, initial conditions, the number and behaviour of agents, and exogenous "world" developments such as the energy price. Rather than shying away from this complexity and only running the model for a few parameter sets (i.e. scenarios), in exploring the possible operation and evolution of socio-technical systems we follow an approach in which once an agent-based model is completed it is run many times to systematically explore the parameter space.

In short, agent-based models are a means towards better decision-making. They allow us to develop models and simulations that allow ex-ante exploration of the consequences of decisions, explicitly recognising that real-world systems are made up of a complex myriad of autonomous but interconnected elements. Agent-based models help us to understand a system's complexity.

1.5 A Book in Two Parts

Agent-based models have already successfully been designed and implemented for various socio-technical systems, including energy and industrial networks, as demonstrated in this book. The book offers a theoretical framework and a practical approach for (agent-based) modelling and simulation of socio-technical systems. It provides recipes for model implementation and guidelines for model development when faced with management and policy issues. It is divided into two parts: the first part gives the background and the theory of the agent-based modelling approach; the second part presents a variety of case studies that illustrate how the agent-based modelling approach has been and can be used. To this end, the case studies presented concern different application domains and different time scales of decision making. These range from short-term decision making at the operational level to tactical and strategic long-term decisions.

We have found the agent-based modelling approach and the case studies presented to be of invaluable support in understanding and steering complex socio-technical systems, and we hope that they will be equally valuable to the reader.

Acknowledgements The contributions to this book stem from researchers in the Energy and Industry Group at the Faculty of Technology, Policy and Management, at the Delft University of Technology in the Netherlands. Many of their projects were executed as parts of the sub-programmes "Understanding Complex Systems" and "Intelligent Infrastructures" within the framework of the Next Generation Infrastructures (NGInfra) Foundation.[2] NGInfra represents an international consortium of public and private organisations working in a concerted research and development effort to improve the performance of critical infrastructures in the face of disturbances, changing economic conditions, emerging technologies, changing public values and end-user demands. Central to the scientific mission of the Next Generation Infrastructures programme is the development of a generic framework, methods and tools for understanding and steering the behaviour of network bound infrastructures, and enabling cross-sectoral learning, with the ultimate goal being the following:

> "To stimulate and support the development of more flexible, more reliable and more intelligent infrastructures and services, with respect for public values and consumer interests, to better serve society in the future".

References

Allenby, B. (2006). The ontologies of industrial ecology? *Progress in Industrial Ecology, an International Journal*, *3*(1), 28–40.

Crutzen, P., & Stoermer, E. (2000). The Anthropocene. *Global Change Newsletter*, *41*(1), 17–18.

de Bruijn, H., & Herder, P. (2009). System and actor perspectives on sociotechnical systems. *IEEE Transactions on Systems, Man and Cybernetics. Part A. Systems and Humans*, *39*(5), 981–992.

Dijkema, G. P. J., & Basson, L. (2009). Complexity and industrial ecology foundations for a transformation from analysis to action. *Journal of Industrial Ecology*, *13*(2), 157–164.

Holland, J. (1992). *Adaptation in natural and artificial systems*. Cambridge: MIT Press.

---

[2] http://www.nextgenerationinfrastructures.eu/.

Kay, J. (2002). On complexity theory, exergy and industrial ecology: some implications for construction ecology. In C. Kibert, J. Sendzimir, & B. Guy (Eds.), *Construction ecology: nature as the basis for green buildings* (pp. 72–107). London: Spon.

Lukszo, Z., Ferreira, L., & Vrancken, J. (2006). Decision making in transport infrastructures. In *Proceedings of the 2006 IEEE international conference on systems, man and cybernetics*, Taipei, Taiwan.

Nikolic, I. (2009). *Co-evolutionary process for modelling large scale socio-technical systems evolution*. PhD thesis, Delft University of Technology, the Netherlands.

Rohilla Shalizi, C. (2006). Methods and techniques of complex systems science: an overview. In *Complex systems science in biomedicine* (pp. 33–114). Berlin: Springer.

Snelder, M. (2010). *Designing robust road networks: a general design method applied to the Netherlands*. PhD thesis, Delft University of Technology, the Netherlands.

van Dam, K. H. (2009). *Capturing socio-technical systems with agent-based modelling*. PhD thesis, Delft University of Technology, Delft, the Netherlands.

Part I
Theory and Practice

Chapter 2
Theory

I. Nikolic and J. Kasmire

Abstract This chapter introduces and explains the main concepts that provide the theoretical background on how to model the ubiquitous socio-technical systems that are so important to modern life. First the notions of systems, adaptation and complexity are discussed as individual concepts before addressing complex adaptive systems as a whole. This is followed by a discussion on generative science and agent-based modelling, with special attention paid to how these concepts relate to socio-technical systems. Throughout the text examples of how the theories can be applied to real systems are provided. Armed with a solid understanding of concepts such as observer-dependence, evolution, intractability, emergence and self-organisation, the reader will have the right foundation for moving on to the practical aspects of building and using agent-based models for decision support in socio-technical systems.

2.1 Introduction

As discussed in Chap. 1, this book is about creating agent-based models to improve our decision making in and around socio-technical systems.[1] Agent-based perspectives, on both analysis and decision support, imply a complex adaptive systems approach to systems and modelling, and this chapter lays the necessary theoretical foundations for the concepts and tools needed to create such models.

The main theoretical perspectives used in this chapter, and throughout the book, are complexity, complex adaptive systems and the generative science paradigm. These perspectives can be considered as meta-theories, which could apply to just

[1] These systems have also been called large-scale socio-technical systems, complex socio-technical systems (Bonen 1981), socio-technical systems (Geels 2004), large technical systems (Bijker et al. 1987), complex innovation systems (Katz 2006), complex engineering systems (Ottens et al. 2006) and even the impressive sounding "system of systems" (DeLaurentis and Crossley 2005). For a detailed discussion on various application fields and uses of complex adaptive systems, please refer to the work of van der Lei et al. (2009).

I. Nikolic (✉)
http://www.igornicolic.com
e-mail: i.nikolic@tudelft.nl

K.H. van Dam et al. (eds.), *Agent-Based Modelling of Socio-Technical Systems*, Agent-Based Social Systems 9,
DOI 10.1007/978-94-007-4933-7_2, © Springer Science+Business Media Dordrecht 2013

about anything, from ecosystems, multinationals, the Internet, the legal system, earthquakes, slang, ant colonies, the brain, immune systems and the pattern of tagging in social networks. However, we use them to explore the domain of socio-technical systems because these systems abound in human society, are fascinatingly complex and unpredictable, are important to the development and welfare of humanity and life on earth, and have so far defied attempts at analysis using other approaches. This is not to say that these are the only valid tools to analyze these systems, or that these approaches are not valid for investigating other systems, but those investigations will not be found in this book.

2.1.1 Focus

Socio-technical systems, as a class of complex adaptive systems (Kay 2002), consist of many technical artifacts (machines, factories, pipelines, wires, etc.) and social entities (individuals, companies, governments, organisations, institutions, etc.). These are interwoven in networks of physical and social components, applying and responding to selection pressures and racing to adapt in a fast vast mega-coupled fitness landscape. Examples include single production plants, regional industrial clusters, interconnected power grids, multi-modal transport networks, telecommunication networks and global enterprises.

Industrial society as a whole is a collection of co-evolving, large-scale, social and technical systems, with multi-dimensional flows of physical matter and social information. Humans, as both physical and social beings, bridge the gaps between the physical networks, governed by laws of nature, and the social networks, with an added layer of social conventions, formal laws, rules and institutions. The social connections are as real as physical ones, varying in length and formality (everything from an interaction with a clerk in store to governmental regulations) (Williamson 1987). In this way, information, research, investment, consumption, institutions, rules, regulations, policies, habits and social values flow between people and physically affect how mass and energy are allocated and used, which in turn shapes the social systems that depend on those physical realities. For example, mixers, reactors, and heat exchangers of a chemical processing plant exchange (and conserve) mass and energy (Coulson and Richardson 1999), but through indirect connections to global supply networks, also influence the price of goods and labour, demand for products, and the stability of governments along the supply chain. Thus, system content, structure and boundaries shift and evolve without any global or central *coordinator*, order and regularity, instead emerging from widely distributed bottom-up interactions of subsystems, some with centralised control and others fully distributed.

2.1.2 Structure of the Chapter

We begin by discussing what systems are. Only after we understand what a system is can we move on to notions of adaptiveness, and how it relates to systems, before alighting on the heady topic of complexity and linking them all for a complex adaptive systems look at some real world examples. Finally, we wrap the chapter up with an introduction to the basics of generative science and agent-based modelling as a tool for a generative science approach to complex adaptive socio-technical systems.

2.1.3 Example: Westland Greenhouse Cluster

Throughout this chapter, we will use the Westland greenhouse cluster in the Netherlands as a running example of a socio-technical system in order to clarify the theoretical notions. Greenhouses represent a good example to work with as they have clear human and social components as well as technical components, yet both are entwined in such a way that there is little value in trying to examine either in isolation.

Modern greenhouses derive from various ancient techniques to alter the environment of plants so as to protect delicate species or to extend the fruiting or flowering season of popular plants, documented as early as fifth-century BC Greece (Hix 1996). These techniques included heating the soil or air around the plants, placing protective materials around the plants, or moving potted plants in and out of protected spaces to best catch sunlight while avoiding extreme conditions. These rudimentary techniques have since developed into sophisticated heating, lighting, aeration, and irrigation systems, as well as various configurations of protective walls and windows, and technologies as modern as computer controlled robotic systems to move or turn the plants, sow seeds or pick produce.

Before the Industrial Revolution, only aristocracy could afford the time, energy and resources for protected cultivation. As such, greenhouses were status symbols and owners competed to have the most diverse collections, best examples of rare species, or impressive new ways of operating the greenhouses (van den Muijzenberg 1980). This competition led to many advances in design and technique that spread and developed through academic horticultural societies, journals, books, as well as through the hiring of experienced gardeners and architects (Hix 1996). Following the Industrial Revolution, materials and technologies became available in quantities and prices that allowed commercial enterprises to incorporate some protected cultivation methods, which brought a new focus on quantity and production lacking in the non-commercial origins of greenhouses (van den Muijzenberg 1980). An a consequence of commercial competition, modern greenhouse horticulture businesses, like those in the Westland greenhouse cluster, specialise in a single product, or an extremely limited range of compatible products, with enormous investments in the tools, technologies and processes designed to optimise production (Hietbrink et al. 2008). This is a highly successful greenhouse area, one of the largest areas of greenhouses in the world, and has a highly technologically advanced processes that use large amounts of resources and contribute significantly to the GDP of the Netherlands. It has spawned equally specialised and high-tech transport, processing, and packaging industries, as well as complicated markets, regulations and subsidy schemes.

2.2 Systems

> "A System is a set of variables sufficiently isolated to stay discussable while we discuss it."
>
> W. Ross Ashby, cited in Ryan (2008)

We open the discussion on systems with Ashby's quote, which introduces several important concepts, some of which are implied rather than stated. Systems are part of the larger whole. They are coherent entities over some period of time. They have boundaries. More subtly, this quote also suggests that a system is something an observer chooses to observe. All of these concepts, and a few more, will be discussed in this section as we explore the history of systems thinking, the contribution of the systems perspective, and elaborate tricky systems notions such as system boundaries, context and nestedness.

2.2.1 History of Systems Thinking

Prior to 1950s, the analytic, deterministic and mechanistic world view prevailed in science (Ryan 2008). People seemed comforted by the idea that everything was ticking over like clockwork, operating under immutable rules, and that given enough time, clever people would be able to measure all the gears, time the cycles, link up the actions and understand how it all worked. There was an answer to everything if only we could see enough detail. This was in no small way linked to the closed system approximations of physics set out in Newton's laws of motion. Physics and chemistry seemed so orderly, so law abiding, so predictable. You could know the future exactly, providing you start with all the right measurements and formulae, and of course have helpful idealisations.[2] Modelling something as a closed system makes it very easy to calculate what will happen because the exact number and properties of all interacting elements is known.

This mechanistic view led to the idea that anything at all could be made finite and fully knowable by drawing some boundaries, learning everything there was to learn about everything inside those boundaries, and then expanding the boundaries to repeat the process. Physics and chemistry pervade the rest of existence, so it only stood to reason that the same orderly, law abiding behaviour that applies when we draw the boundaries around atoms and molecules must apply to everything made from those atoms and molecules. At higher levels, the levels of plants, animals, people, fashions, stock markets and technological development, there are regularities

[2] Please be careful, we seem to have misplaced our frictionless surface.

and patterns that bolstered the belief the rules governing these too could be made obvious and written down. And what hope! We could sweep away all unfairness, poverty, waste, and bad music, among other evils, if only we could find out Newton's three laws of humanity! All of science strove to prove that other disciplines were as well behaved and tractable as physics and chemistry, and that, in time, we could understand it all.

Yet differences were observed between the mechanistic predictions of society based on the behaviour of atoms and the reality of society, and these discrepancies were increasingly difficult to justify as errors in calculation or measurement. The idealisations began to look impossibly distant and systems somehow never seemed to be as closed as they needed to be. Exceptions were found to the laws governing physics, although not on the scale of ordinary life. Self-organising chemistry was discovered, which, while obeying all the rules, remained stubbornly unpredictable in the big picture. And all the rules found for disciplines like biology, psychology, and sociology seem to come with a list of exceptions that makes them almost useless.

Slowly, the scientific community came to appreciate that perhaps law abiding behaviour at low levels did not equate with predictability and control at higher levels. The idealisations and falsely closed systems of physics and chemistry just didn't scale up. Something else was going on, and that turned out to be that the kinds of things people wanted to study were often open systems, where matter and energy flow in and out, and where things inside a system are affected by the environment outside the system. This changed the view of systems from a boundary around all things that interact to a boundary around those things that interact together in such a way that makes them interesting enough to draw a boundary around. Among all of this, the development of General Systems Theory (Von Bertalanffy 1972), strove to argue that open systems were valid scientific study and that a system is not a matter of absolute truth but a matter of point of view, of usefulness for a purpose, and of relative value. The system genie had been let out of the bottle.

Ever since people have been struggling to reconcile the leftover desire for an predictable and eternal clockwork world and the unpredictable world that only looks like clockwork when you chop off everything interesting. This had led to quite a lot of interest in ways to talk about, study, control and influence systems, and to capture the mysterious "what else" that is going on between the movement of atoms and the madness of crowds.

Greenhouse Example Let's consider how the greenhouse horticultural sector might look from a mechanistic perspective, with all the idealisations and assumptions that it entails.

Each greenhouse would be an optimised converter of raw resources (nutrients, seeds, energy, etc.) to finished products (vegetables, flowers, etc.). The basic conversion process would follow an immutable formula, so that a fixed amount of resources would always produce the calculated amount of product. As each greenhouse is optimised, there would be no reason to believe that the greenhouses are not equivalent or interchangeable, or that the production or efficiency of conversion might change over time if not deliberately changed.

The flows of resources and products would be clear and unambiguous, and the flow of information between all components, such as growers and suppliers, would be perfect and synchronised. Fully informed rational decision makers would always behave identically, so predictions are relatively simple.

Regional governments would control the behaviour of the sector by regulating changes that put in or remove barriers to the flow of resources or products or by altering the conversion formula with taxes, subsidies or the like. These controls would achieve maximum competitiveness in the global markets. Any change in the system structure is assumed to be deliberate, rational and coordinated.

While charming, this account of a regional greenhouse horticultural sector looks more like a game of poker where all players have identical hands, all cards are played face up, and with an arbiter controlling the rules to achieve perfectly distributed bets. Clearly, this does not represent what we observe in reality. Greenhouses are not formulaic black boxes that convert x amount of seeds and water into y amount of tomatoes, nor are the inputs fixed or predictable. Participants are not fully informed, are far less than rational, and behave unpredictably. Regional governments find that their efforts to control systems are ineffective or produce the opposite of the desired effect, changes to the system are uncoordinated and the last thing from deliberate, markets are far from free and power is very unequally distributed along the supply chains.

The systems perspective, which has largely superseded the mechanistic world view, is far more useful here. But what is that system perspective? Follow us into the next section to learn more!

2.2.2 Systems[3]

Systems are many things to many people. To be clear, although the closed systems of Newtonian physics are also systems, we want to talk about the systems with fluid edges, boundary crossing influences, and inexplicably organised internal elements that none the less confound our attempts to describe them.

If we look in a dictionary, we find several related but independent uses of the word "system":[4]

- A regularly interacting or interdependent group of items forming a unified whole.
- An organised set of doctrines, ideas, or principles usually intended to explain the arrangement or working of a systematic whole.
- Manner of classifying, symbolising, or schematising.
- Harmonious arrangement or pattern or order.

[3] Recursive structure is a characteristic of systems.

[4] http://www.merriam-webster.com/dictionary/system.

For our purpose of understanding and managing the evolution of socio-technical systems, the first and last definitions are useful. Systems consist of interacting and interrelated elements that act as a whole, where some pattern or order is to be discerned. But we can only say they act as a whole because we see a pattern, or perhaps, we only see a pattern because we define the system as acting as a whole. If we saw a different pattern, we might define the system as a slightly bigger or smaller whole, or if we had already defined a different whole system, we might see a different pattern. The circularity of this definition can be a little boggling, so we will adapt a definition from Ryan (2008). Systems:

1. are an idealisation;
2. have multiple components;
3. components are interdependent;
4. are organised;
5. have emergent properties;
6. have a boundary;
7. are enduring;
8. effect and are affected by their environment;
9. exhibit feedback; and
10. have non-trivial behaviour.

We are getting closer, but these points need a little more elaboration to be sure we all agree on what systems are.

Idealisation Systems are not actual entities, as such, but are idealisations or abstractions of a part of the real world. A given system might seem clear enough, an obvious unit or entity that apparently stands alone, like a greenhouse. But closer inspection reveals that no two greenhouses are exactly alike, with, for example, some producing their own energy while others buy from the energy network. Thus, to get a useful abstraction or idealisation of a greenhouse system, we must either exclude the energy production facilities of some greenhouses or ignore those that buy their energy.

Multiple Components Systems always consist of multiple components, usually guided by the structure of a system. Individual greenhouses might be components in the greenhouse regional cluster system, even while the greenhouse system is composed of individual technology components, like combined heat and power (CHP) units or aeration units.

Components Are Interdependent Systems differ from unorganised *heaps*, which also have multiple components, by the fact that the elements are interdependent and interact. For example, lighting systems can add heat to the greenhouse, even though they are not a heating systems, and both lighting and heating systems use power, which might be generated by a CHP unit.

Organised The interaction and interdependence is not random and unstructured, but follows a certain pattern of interaction. While it is not impossible for each component of a system to interact with all other components, in most systems certain components interact more tightly with a subset of components, and the interactions are of a limited type or direction. For example, tomato growers participate in tomato growers associations, and rarely interact with flower growers, even if they are their direct physical neighbours. The interactions are usually limited to discussions, demonstrations and the like, and typically exclude such possible interactions as backrubs, fight clubs or paid musical performances. Within the limits of the associations, some kinds of interaction are bidirectional, such as discussions, while others, like votes or announcements, are unidirectional, further organising the structure of the interdependence.

Emergent Properties As will be discussed in more detail in Sect. 2.5.2, complex systems display properties that cannot be understood by just looking at the properties of the individual components, but are created as a result of the structure and organised interactions between these components. For example, the price of tomatoes can not be directly determined by just looking at the costs of the facilities and resources used to grow them, nor even by looking at the total production and demand. The price of a tomato is determined by many things, not all of which are operating at the level of tomato production.

Boundaries Every system description must contain an explicit definition of what is in the system and what is outside it. This is relatively easy for closed systems or highly idealised systems, but more realistic attempts to describe social and technical systems quickly find that drawing these boundaries is tricky business. The decisions on what to keep in or out of the system description depend on who is looking at the system, what they observe, and with what purpose they are making this observation. For example, when describing the boundaries of the horticultural system in the Westland, a politically based system might prefer to draw a strict geographical boundary at the edges of the municipality of Westland, while an economically minded description would likely include parts of the network of economic interactions around the greenhouses, even if they were located over the municipality border.

Enduring A system can only be considered a system if it lasts long enough to be observed or discussed, but conversely, it can be so enduring that it is hard to observe or discuss. Close observers will measure and study systems that others dismiss as too transient, but might fail to notice very slow acting systems with changes too long-term to be readily noticeable. For example, we easily consider the horticulture sector as a system because it has existed as a clearly identifiable kind of agriculture for a sufficiently long time that people have studied it as a system. On the other hand, an individual horticultural company that goes bust quickly would be too short term to interest most researchers, and although agriculture itself is a system, it is so long standing and wide spread that observations or discussions of agriculture as a whole are almost meaningless.

Environment Defining a system as an observer-dependent abstraction with observer-dependent boundaries means that the system is a particular interpretation of a particular subset of the real world. The "rest" of the real world is the environment in which the system is situated. All systems communicate mass, energy and information with the environment, so are to some degree "open", but to simplify the system description, we abstract the environment into only those variables or parameters that are most relevant for the system. For example, when understanding the Westland greenhouse system, we might consider Chinese and US tomato growers as an influence from the environment, while we would probably not consider a music festival in Rotterdam, even though it is right next door, as music festival visitors are unlikely to alter the tomato system to a noticeable degree. However, if it became fashionable to buy a kilo of tomatoes after attending such a music festival, then we would need to include the festival in the environment of the tomato growers.

Feedback The interactions between system components are not only organised, but also contain loops, where A influences B and B influences A in turn. These loops create feed-back and feed-forward mechanisms that give rise to non-trivial behaviour. A positive feed-back loop, where the success of something drives more success, would be exemplified by the introduction of the "Tasty Tom" tomato variety, which was popular enough to generate high demand, driving more production, resulting in access into new markets, where it also proves popular and drives yet more demand. A feed-forward loop example would be the case where an anticipated drop in product prices might encourage growers to maximise quantity of production rather than quality, which results in a market flooded with poor quality products that receive a lower price, driving producers to carry on maximising quantity rather than risk switching to quality focused production.

Non-trivial Behaviour Foerster (1972) describes trivial behaviour as invariant mapping between system inputs and outputs, but the different feed-back, feed-forward and other interaction loops, coupled to inputs from the environment create non-trivial behaviour in complex systems. For example, if the price of tomatoes drops sharply, a trivial behaviour would be a proportional drop in supply, but in reality the supply does not behave trivially. Instead, although supply might be affected, tomato growers must balance the drop in price with long-term contractual agreements with suppliers, a sense of pride or family traditions in producing tomatoes, significant investments that need to be recouped, and many other factors, irregardless of the economic conditions.

2.2.3 World Views

Whenever we interact with an system, we are never passive or objective observers. In order to see, we need to choose to look, and although reality is infinite, our powers of observation are limited. To cope with the onslaught of infinite reality,[5] we have developed several strategies to whittle reality down to comprehensible observations. We observe everything through the lens of our human nature as well as that of our individual world view, which is not entirely under our own control. At the same time we are, to some degree, in control of our powers of attention and focus, or what it is that we choose to observe and how we do so. The question therefore is not whether there are different world views, but how these world views affect the observations and interpretations of the system it views.

Some of these universal strategies to gainfully reduce the tide of the potentially observable are derived from evolutionary history. We preferentially attend to rhythmic motion because it is generally more urgent to identify animals than plants, and we cheerfully ignore microscopic organisms because our human scale is ill suited to that size. But other strategies are part of the world view, or fundamental and relatively consistent cognitive orientation of an individual or society, and include biases that are not necessarily shared, but are derived from experiences and interests. Classically trained musicians will notice off key notes that most would not, while engineers will be fascinated by mechanical devices that musicians might ignore. These world views determine not only what we pay attention to, but also provide a framework for generating, sustaining, and applying knowledge. For example, economic woes will appear to be the direct result of too much or too little regulation depending on how financially conservative the observer is, with each side arguing that any evidence against their point of view must be faulty.

Of course, both sides could be said to be right because they are essentially disagreeing about how they define the system. No two observers will agree on system boundaries, how to parse the system into relevant parts, or how to rank, measure and track the things determined to be of interest, all of which in turn will determine what we attend to, measure and observe in the future. This is especially apparent with socio-technical systems, as there are no unambiguous boundaries, no natural scales, and few shared contexts to guide our attention, efforts and observations.

Greenhouse Example As we are not greenhouse farmers, we find that we have a very different perception of technology than they do. For example, when deciding on the purchase of a new technology, greenhouse farmers seem to view technology as unique and personal, for which only personal experience can be of help. They tend to value their own experience the most, followed by the experience of other greenhouse growers. They express a distrust of the information provided by the technology producers, academic analysis, or government agencies, perhaps expecting it to be painted in a favourable light or applying only abstractly to the actual performance that can be expected. Essentially, the less the person giving the information is like the grower, the less they trust the information given.

But as academics, we are inclined to view technologies as homogeneous and impersonal, and obviously do not see a problem with trusting the results of scientific

[5] It just keeps coming.

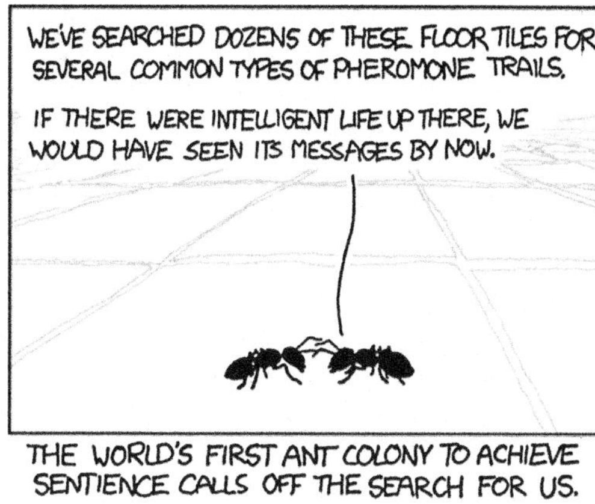

Fig. 2.1 Observer-dependence. By Randall Munroe (http://xkcd.com/638/), used with permission

analyses, especially if the data is provided, expecting the analysis to be replicable and representative of the technology performance. Yet academics might also express a distrust of corporate figures, government statements and the anecdotal evidence of greenhouse growers, if there is no scientific data as support. Which just goes to show that greenhouse growers might trust themselves first and those most like themselves second because the information is so specific and individual, but academics do the same because they believe the methods of science eliminate the specificity or individuality of the subject. Perhaps growers and scientists are not so different, despite the orientation of the distrust.

2.2.4 Observer-Dependence

While world view affects all aspects of perception and understanding internal to the individual, observer-dependence affects the externalisation of that world view as it applies to a system under study. Every observer must choose the scale at which the system is observed, including time scales, and thereby determines what is considered the smallest/largest elements or periods of interest. For example, one observer might see people as agents, interacting in a legal system over the course of months or years while another observer sees entire countries as agents, engaged in worldwide political games over a matter of decades. Furthermore, each observer chooses a certain perspective when interacting with a system. For example, a chair can be viewed as an artistic, social, mechanical or chemical entity which determines whether the chair is seen as an example of post modernism, a variable leading to improved classroom behaviour, or as an object defined by properties such as mass, strength and volume. The cartoon depicted in Fig. 2.1 illustrates how one's choice of what and where to observe can support an erroneous idea.

Objectivity

> "An objective account is one which attempts to capture the nature of the object studied in a way that does not depend on any features of the particular subject who studies it. The object

> has certain properties or behaves in a certain way even if the subject (you as a person) does not see it. An objective account is, in this sense, impartial, one which could ideally be accepted by any subject, because it does not draw on any assumptions, prejudices, or values of particular subjects. This feature of objective accounts means that disputes can be contained to the object studied."
>
> Gaukroger (2001)

Science is at least implicitly understood to be a method for exposing irrefutable facts, and objective, real truth. Those who disagree with the findings of science, like creationists, do not usually argue that science cannot find the true nature of things, such as mechanisms for cell growth or immune system functioning. Instead they suggest that a particular application of science is faulty or biased in some respect, or that there are aspects of life that science cannot examine for truth, like the ineffable nature of the creator, even if such an investigation can find truthiness in fossil records or geological studies (Colletta 2009; Munger 2008). In effect, almost all of us believe that science is capable of objectivity, although not universally so.

But the concept that science can truly be objective is an unattainable goal, an unhelpful ideal, and a needlessly divisive crutch, even for the most uncontroversial topics or objects of study. Objectivity, seems so attainable, but is elusive to the last because the observer is responsible for selecting the object of study, the aspects to measure or record, the tools or instruments to use, and the methodology to follow, among other crucial choices that determine the final outcomes of the scientific endeavour. Inevitably, some features or qualities that could be measured will be ignored or mis-recorded, and the instruments and methodologies chosen cannot help but be less than perfectly precise or balanced. Furthermore, the interpretation of the data collected, the creation of models to best fit that data, and the predictions of the future, which guide measurements and research to come, are also highly dependent on the scientist rather than the object of study.

Total objectivity is arguably not possible in some, or maybe even all, situations. But do not despair! While science cannot promise objectivity, we can at least be aware that:

- each observer has its own world view that determines what and how observations are made;
- observer-dependence interacts with emergence; and
- observer-dependence affects the process of model creation.

We can not reliably and objectively determine what effect these will have, but awareness of these issues makes one better equipped to approach observer-dependence. For a more controversial approach to objectivity in science, please refer to the work on Post Normal Science by Funtowicz and Ravetz (1993).

Greenhouse Example Subjective analyses come from very different sources. For example, two successful greenhouse growers looking to purchase the same property in order to increase their production will offer different bids. They will have taken different aspects of the same property into consideration and arrived at different subjective evaluations. For example, one will take the relatively high energy consumption to be more important than the benefits of an excellent location near an auction house, while the other will value the location highly and dismiss the importance of the current energy demands.

Another, and far more subtle, example of objectivity vs subjectivity would be the measurement of the temperature in the greenhouse. Decisions must be made as to whether the temperature should be measured in one or more locations, where these locations are, how often the measurements are made, what time of day they are made, and the type and sensitivity of the device employed to measure and record the temperature. While we might think that temperature is objective, there is no really objective way to measure it, and the choices made will influence the data gathered.

Reductionism and Holism

> "The utility of the systems perspective is the ability to conduct analysis without reducing the study of a system to the study of the parts in isolation."
>
> Ryan (2008)

Reductionism is the idea that system behaviour is determined by the behaviour of the system components, and is best exemplified by the idea that a thing is nothing more than the sum of its parts. Reductionism can be understood as an attempt to achieve objectivity, by reducing the systems to smaller, more observable, and therefore more objective, pieces. According to reductionism, understanding all of the parts is equivalent to understanding the thing. Related to this idea, downward causation says that the upper levels of hierarchical structures constrain the actions of the lower levels, or that to understand all of the parts of a thing, you need only understand the thing. Regardless of whether a thing can be understood only in terms of its parts, or the parts can only be understood in terms of the thing, both reductionism and downward causation quickly lose meaning in nested systems, where each of the parts are in turn made up of sub-parts. Taken too far, this leads to large-scale behaviours or events, like consciousness or stock market crashes, being explained in terms of the laws governing sub atomic particles.

The opposite stance would be holism, where a system, be it physical, social or biological cannot be determined or explained by its component parts alone. Put more colloquially, the whole is more than the sum of its parts. This view is linked to upward causation, where the parts that make up a system are not constrained in any way, but the system is constrained by those parts, so that the lower level system components provide all the possible behaviours that a system can have. Like reductionism, holism can be taken to ridiculous lengths by claims that it makes no sense to examine the component parts of a system at all and that only a study of the system in its entirety leads to understanding.

The kind of reductionism that absurdly tries to explain any phenomenon based on the smallest possible components is called greedy reductionism (Dennet 1996), and while the explanations of the smaller components are not necessarily untrue, they are remarkably unhelpful for explaining the higher level system. Likewise, extremely holistic approaches are unlikely to be of any use, especially for addressing urgent or current issues because an exhaustive study of whole systems is time consuming and difficult, if not impossible, and extremely inaccessible to anyone who has not studied the same system.

More fruitfully, the extremes can be avoided by recognising that the links between the thing and its parts are influential but not completely causal. If the behaviour of the system and its constituent parts influence or constrain, but do not completely determine each other, then there is a clear benefit to looking inside and observing the parts and interactions while also looking at the higher levels as well. Yet avoiding extremes of reductionism and holism demands an understanding of other problems, such as the observer-dependence of delimiting the boundaries or defining the contexts of the systems under study.

Greenhouse Example A reductionist stance would be that the performance of a greenhouse is only a function of the technologies that are in it. An extreme holistic stance would claim that each greenhouse has a unique performance, and it irrelevant that all top performing greenhouses are all using a particular technology for heating. A useful reductionist approach will perceive the connection between the performance of the greenhouses with the technologies implemented within in, but will also recogniser that the location, technology interactions, state of maintenance, management style of the owner, as well as factors outside of the greenhouse, are also relevant.

2.2.5 System Boundaries

> "The real challenge posed by the systems idea: its message is not that in order to be rational we need to be omniscient but, rather, that we must learn to deal critically with he fact that we never are."
>
> Ulrich (1988)

There is no system with an outer system boundary through which no energy, matter or information penetrates to influence the internal workings of the system, although some systems are simple enough to be usefully modelled as if such boundaries existed. The systems of interest to us, however, are not so easily idealised, so we must decide which parts, relations or influencing factors are not known or suspected to influence the system strongly enough to be worth the effort of including. And we must do so in full awareness that an arbitrary boundary is drawn around what seems like a useful subset of a larger system (or systems) when we observe a cluster of activity that warrants further study.

Although in reality, everything influences everything else in some way, Ulrich is right to stress that we do not need to know how everything is connected. We really only need to be aware that the boundaries we decide to set will reflect our needs and goals rather than the true nature. If we want to examine how rising fuel prices affect the greenhouse horticulture sector in the Netherlands, we might or might not want to include representations of transport, horticulture and agriculture in other countries, technological development for energy efficiency or international markets. All of these things, and many more, will be connected to the greenhouse sector, so there is no inarguable system boundary to be drawn. But if too many connections are included then clear relationships and influence will be harder to elucidate and the model will be no more enlightening than simple observations of the real world would have been.

2.2.6 System Nestedness

When the highest level of one system is also the lowest level of a larger system, then the systems are vertically nested. Each system can be viewed as an isolated whole, or can be viewed as composed of other systems and residing in an environment made of the next higher level system. Deciding whether a given level will be represented as a unit or as composed of smaller units can be understood as a sort of vertical system boundary. The level of observation must therefore also be made without pretense that it is the only possible level at which the system could be observed. Instead, it is the level at which the expected observations are most likely to lead to improved understanding of a given question about the system.

Although conceptually arranged in hierarchies, Hollings points out that these "hierarchies" are not top-down sequences of authoritative control, but rather, semi-autonomous levels formed from the interactions among a set of variables that share similar speeds (and, we would add, geometric/spatial attributes)" (Holling 2001). These nested or hierarchical arrangements are a sort of conceptual shorthand based on the way evolution tends to develop stability from frequent interactions, both in time and space. These stable interaction patterns appear as structured units which interact with similar units and serve as building blocks in larger structured interactions. For example, the inhabitants of a town interact much more frequently with each other than they do with the inhabitants of another town, so each town could be considered a system, embedded in a larger system for the region or country. But even though the towns are usefully idealised as separate systems, the residents of each town are not constrained or prevented from interacting with the other, they are just less likely to do so. Instead, in nested systems, the subsystems overlap, and it is this overlap and the interaction that it enables that given rise to complex behaviour Alexander (1973).

Because the lower level stable structures are much more likely to interact on the same scale, the interactions at other levels seem remote and simplified, the more so the more distant the level. Therefore, "three levels of a hierarchy, one up and one down, are generally considered sufficient for analysis of the focal level" (Ryan 2008). Although, if there is any disagreement as to what is the focal level, there will of course be disagreement as to what the upper and lower bound of observation should be.

It is important to note that systems can be nested in time, physical space and social relations, among other possible ways, and that every system can belong to more than one larger system as well as be composed of more than one arrangement of smaller systems.

Greenhouse Example A greenhouse farmer can belong to a physical neighbourhood, inside of a town, inside of a district, inside of a country, as well as belonging to more than one growing association, inside of larger trade unions or industrial sectors. The farmer is also likely to belong to family units, inside of a larger extended family as well, and to a local religious group, club or team inside of larger associations. Inside the greenhouse, systems can also be nested. A temperature regulation feedback loop is a system nested within the greenhouse heating system. The heating system is nested within the climate control system of the greenhouses, together with the aeration, lightning and irrigation system. The greenhouse is a system nested within the district heating system and power grid of the region of Westland, which in turn are nested within the Dutch and ultimately European power grid.

2.3 Adaptive

To be adaptive is to have the property of adaptation, or improvement over time in relation to environment. The environment need not be physical, as social, technical, and cultural environments can also cause adaptations.[6] Adaptation is not the same as change in response to a stimulus because adaptations are specific kinds of changes in response to specific types of stimuli. The changes must be improvements (how to determine what is an improvement is covered in the next section) as changes that make something worse or merely different while being neutral in respect to the environment are not adaptations.

Further, the changes must be in response to stimuli from the habitat or environment. These stimuli can be purely environmental, like temperature, terrain, or the availability of resources to which adaptive entities can be come better suited. The stimuli can also be contact or interaction with the great diversity of other entities, ranging from direct adversarial interactions, such as predation, competition for resources, and parasitism, to beneficial interactions such as cooperation and symbiosis.

Lastly, the stimuli from the habitat must be constant or reoccurring. Gravity is static, and of course everything that has adapted has adapted to deal with gravity. Likewise, entities can adapt to the dynamic but periodic (within a lifetime) patterns such as climate and diurnal rhythms, growth and decline of population numbers, cycles of resource availability, seasonal migrations, and many others. These relatively predictable and ordered environmental stimuli can force adaptations, but

[6]The specifically improved features, behaviours or traits are also called adaptations, but it is preferable to refer to these as adaptive traits to avoid ambiguity.

catastrophic change events cannot. Some events are so disruptive that a great majority of species go extinct, areas become uninhabitable or unusable, and built environments are completed destroyed. Nothing can adapt to such extremely rare, sudden and utterly devastating global catastrophes.

Adaptations are not just change, or even change in response to stimuli, but neither is adaptation the same as evolution.

2.3.1 *Adaptation Versus Evolution*

While adaptations are improvements in response to environments, evolution is the algorithmic process that produces these improvements, best summed up by the famous maxim: "Vary, multiply, let the strongest live and the weakest die" (Darwin 1985). An enormous body of knowledge now exists on the evolution and co-evolution of biological systems following the publication of Darwin's book "On the Origin of Species" (Darwin 1985), which has now developed into one of the best researched and supported scientific theories today. Although Darwin did not understand the specifics of how DNA, mutation, or some selection pressures worked, he quite rightly surmised that evolution will occur whenever certain conditions are met.

The first of those conditions is that there must be differences between things, or variation. Variation occurs, for example, through the addition of totally new material, either through creativity or merely as a result of copying errors, and through the combination of existing designs or genes in new ways.

The second condition is that variations must be replicable or heritable in some way, so that even if offspring or copies are not perfect, they are at least more likely than not to have some of the same variations of the parent or original. This condition of replicability seems to be in conflict with that of variation, with one demanding differences and the other similarities. Of course, the key is that neither is absolute, balancing each other out so that descendants are more similar to the progenitor than to others, without being identical.

The final condition is that there must be some determiner of which replicable variations are better than others. Selection, or selection pressure, is the force in the environment that determines how well suited one variation is to the environment, sifting them mercilessly into the winners and losers, the quick and the dead. Darwin noticed that many animals did not live long enough to reproduce, while some were very successful and had many offspring. Of course we see the same pattern in products, music groups, sports teams, companies, and a myriad of other entities that become successful or not as a matter of the selection pressures acting on them. Only those variations with a leg up on the competition last long enough to produce copies or offspring to populate the future.

Crucially, all three conditions are equally important for the proper functioning of the algorithmic process, and it reveals that evolving things are more than just a collection of matter and chemicals. They are "boring stuff organised in interesting ways" because there is also information, encoded in a the structure of the matter and chemicals, and in the and relationships between these structures and the environment.

2.3.2 Evolution—More than just Biology

Traditionally, evolution has been most readily recognised and commonly accepted in the Darwinian process of natural selection that drives changes in the genetic material of living beings to adapt to their physical habitats. But, as an algorithmic process it is domain neutral and can explain non-genetic changes shaping non-biological adaptations as well. Darwin thought languages were also evolving through the same processes as organisms, and scholars such as Mandeville and Harth (1989), Hume (1962) and Smith (1963) considered that industry, law, politics, economies, markets and manufacturing were also shaped by evolutionary forces. Not having physical matter or DNA as biologically evolving entities do, languages, industries, laws and the many other non-biological entities must use some other system, and the leading theory (Dawkins 1990) suggests that *memes*, cultural, self-replicating entities, analogous to genes, are responsible for the spread, construction and evolution of culture. In the words of Dawkins (1990) and Dennet (1996):

> "A meme is the basic unit of information which spreads by copying from one site to another and obeys, according to Dawkins, the laws of natural selection quite exactly. Meme evolution is not just analogous to genetic evolution. It is the same phenomenon. Cultural evolution simply uses a different unit of transmission evolving in a different medium at a faster rate. Evolution by natural selection occurs wherever conditions of variation, replication and differential 'fitness' exist."

Proposed examples of memes are traditions, technologies, theories, rules and habits. This book can be seen as a meme that might or might not survive over time, depending on its usefulness to the persons who are aware of its contents and find it useful enough to tell others about it.

Ziman argues (David 2000; Jablonka and Ziman 2000) that both biological and socio-technical entities display variation, replication, and selection through a succession of generations, determined by reproduction cycles in organisms, and cycles of learning or imitation in social systems. Further, Ziman observes (David 2000) many phenomena that arise as a consequence of the evolutionary algorithm in both domains, including diversification, speciation, convergence, stasis, evolutionary drift, satisfying fitness, developmental lock, vestiges, niche competition, punctuated equilibria, emergence, extinction, co-evolutionary stable strategies, arms races, ecological interdependence, increasing complexity, self-organisation, unpredictability, path dependency, irreversibility and progress. While there is a lot of support for the fact that social and cultural things are also evolving by the same algorithmic process as biological things, there are several points or criticisms that merit further discussion, although most of these stem from ill-advised attempts to use a strict biological analogy as the model for non-biological evolution.

First, there is no clear relationship for social and cultural artifacts that compares to the organism and gene relationship. This immediately proves a deal-breaker for many critics of memetics, but there are good reasons not to dismiss non-biological evolution out of hand. The relationship between organisms and genes was not at all clear when Darwin proposed his theory, but that didn't stop further investigation. Additionally, the link between genes and organisms is not as well understood as it may seem because of the complex, epistatic relations between genes, preventing any simple one to one cause and effect.

Second, the idea that genes have a special role in evolution only holds for advanced organisms. Not all evolving organisms have DNA, and the very first replicators were not organisms with any kind of genetic code. They were most likely simple self replicating molecules, with no distinction between genotype and phenotype. It is not clear if socio-technical evolution has moved beyond the early stages to arrive at special roles for the non-biological equivalent to the gene.

Third, many researchers cheerfully state, without clear evidence, that technical and social artifacts are not generated randomly but are purposefully designed, although randomness is so important that many designs are described as serendipitous flashes of inspiration (Roberts 1989), accidents that prove useful or the lucky result of arbitrary task assignments during the design process. Further investigation reveals that even those revolutionary, clever designs were built on foundations of rigorous testing, trial and error and incremental advances on previously successful designs, which looks quite a lot like the random mutations in each generation of biological organisms.

Fourth, biological evolution operates in generations that are strictly vertical (no one can be their own parent) while social and cultural evolution has been suggested to move horizontally or reverse vertically, as when someone learns something from a parent, and then teaches that parent an improved version of the lesson. However, this confuses the relationship between the people that teach and learn with the things taught or learned. Children can certainly learn from and teach their peers and parents, but a thing can never be taught without having been learned first. Thus, the things that are actually evolving though non-biological evolution are also strictly vertical, in that they must be learned before being taught, just as organisms must be born before they can reproduce.

Fifth, the speed of evolution is much faster in culture, since there is no need for the genetic transfer taking place over generations. It happens in the "meme sphere", the shared human cultural space, and not in the biosphere. Although entirely true that the speed seems to differ, this can hardly be considered a criticism.

And finally, created artifacts are not alive and do not reproduce in the sense of creating offspring. For some people, aliveness seems to be a necessary condition for evolution, despite the fact that viruses clearly evolve, requiring a new flu jab every year, but are not often considered alive. Lee Cronin, an inorganic chemist, has suggested that anything that evolves should be considered alive, but the opposite was once considered commonly accepted knowledge. Thus, Darwin compared evolution by natural selection to the development and change in languages, because "everyone knew" that non-living things could undergo gradual change processes while living things were immutable.

Although memes appear to be one of the best ways of approaching socio-cultural evolution, they may not be as necessary as many might think. It may be preferable to look at socio-technical evolution on its own rather than through the lens of biological evolution because the systems appear to be two examples of a generic evolutionary algorithm instead of one being real evolution and the other an analogy.

2.3.3 Adaptation in Its Many Forms

Adaptations always start with what is currently available for use, and improve it or apply it in new ways that do not decrease fitness in the immediate term. The best adaptations are those that use the tools at hand the best, not those that can identify what the best of all tools would be. But adaptations are not just the physiological traits of exotic animals that come to mind so readily. There are actually three levels at which different kinds of adaptation can be found, all of which take advantage of different tools and operate on different time scales. These are the individual, cultural and biological scale.

The shortest time scale for adaptation is the individual lifetime, giving us individual adaptiveness. The best example of individual adaptiveness would be learning (Argyris and Schon 1996), although other examples include muscle development in response to constant use, improved immune response after exposure to pathogens, and changes in appearance or behaviour, such as developing a tan in a sunny location and decorator crabs that add bits of seaweed to their shells. In the non-biological realm, individual adaptations would include the unique wear and tear that makes something perform better as it "breaks in", the addition of new words in the personalised T9 dictionaries on mobile phones, or the gradual increase in predictive power of smart systems as they find patterns in their input sensors. Anything that an individual could do, in it's own lifetime, to become better suited to its environment is an individual adaptation, but these are not replicable.

If an individual adaptations comes to be imitated or reproduced as a consequence of learning, then it forms the basis of cultural adaptations. Cultural adaptations take place at a slower time scale than individual adaptation because they necessarily involves at least two occasions of individual adaptation. Humans do many things that fall into this category, so much so that it can be difficult to see that the many ways that adults teach youngsters to stay safe, gain access to food, communicate with others or avoid danger are cultural adaptations. Animals too have cultural adaptations, although less universally agreed upon, these include explicit lessons, as when chimpanzees teach their offspring to break open nuts or angle for termites, and also non-explicit cases of imitation such as songbirds learning to copy the local dialect, foraging habits, food preferences, nesting sites, etc. Social and technical examples of cultural adaptation are also rife, and include schools of thought on everything from economics to the right way to serve tea, designs of houses, models for approaching the design of tools, measuring devices, infrastructures and systems, and just about everything else that people teach or learn not directly related to survival.

The longest time scale for adaptations is also the most well known. Biological adaptations develop over many generations and are not learned, neither individually or culturally, but relate to inborn traits or instinctual behaviours. Biological adaptations can be quite spectacular, such as the angler fish's glowing and twitching lure which helps him attract food in the dark depths of the ocean, or quite unremarkable, such as the proper functioning of internal organs. Not having biology, socio-technical artifacts may not be evolving at this level, although further development, such as in self-replication robots, may require the replacement of "biological" evolution with something that operates at this level and timescale but includes both biological and non-biological evolution.

An adaptation that first appears at one level, can potentially move to another. If an important lesson is learned by an individual, they may teach another or be imitated without explicit teaching, moving the adaptation from the individual to the cultural level. If the environment allows for a Baldwin effect[7] (Weber and Depew 2003), then the lesson may become a genetically inherited instinct, resulting in the same outward behaviour but without the effort or risk involved in the learning process by moving the adaptation from the individual or cultural level to the biological.

2.3.4 Direction of Adaptation

Adaption and selective pressures often appear to "want" to go in a particular direction, toward what is called an attractor, and can display self reinforcing effects that seem to accelerate the direction of adaptation. This can be represented visually through fitness landscapes (see Fig. 2.2), a description of the conceptual environment of an individual or species, where every point in the landscape corresponds to one possibility in the space of possible, with the fitness of each possibility represented as altitude. The more fit possibilities are the attractors, depicted as the peaks of hills. New variations are made by stepping from one location to another, and selection pressures reward those steps that move uphill, or the adaptations, and ignoring lateral steps as neutral changes. Steps that move downhill are punished as deleterious or maladaptive changes.[8]

While an evolving entity located on a plain between hills could go in many directions without changing fitness, once in the basin of attraction for a peak, the only way to adapt is to continue climbing the hill. Progress up the hill then appears to accelerate due to the self reinforcing effects of adaptations, where upward motion means that there is less room for lateral movement. If the hill also grows steeper then each step equals faster movement toward the peak. Step size, or the rate at which things can change, becomes an issue too. Bigger steps means faster progress toward the peak, until the peak is closer than the size of the step, at which point

[7]A theory exploring how acquired or learned behaviour can become integrated into an orgnanism's genetic markup.

[8]Alternatively, the attractors are sometimes depicted as the valleys in the fitness landscape, based on the premise that things can only roll downhill, and do so quite naturally.

overshooting the peak means no progress is possible. On the other hand, very small steps are likely to ultimately get closer to the peak, but will take much longer.

Evolution can only locally optimise so evolvees may get driven up a hill that is not the highest on the landscape, but only coincidentally the first encountered. However, movements that descend the hill to search for a better hill are impossible because even a temporary decrease in fitness will be quickly killed off by the short sightedness of selective pressures. As such, hills can only be climbed, never descended, giving the appearance of inevitability or a teleological drive. This misleading interpretation derives from focusing on the well adapted and successful. Viewing only the victors of ruthless competition suggests that becoming better is somehow the meaning of life, the universe and everything. From a wider perspective, it is clear that there is no goal or purpose, because there are far more losers than winners, and that the competition is endless.

Theoretically, socio-technical evolution allows for long jumps, from the top of an attractor hill to what may prove to be at least equal altitude on the slope of another hill by means of imitating what appears to be a more promising design or solution. However, it is not clear that imitation can really be considered to be "jumping" from one peak to another. It depends on the point of view as to what it is that is walking up the hill. If a company is walking up the hill, then abandoning one product design in favour of copying that of a rival would certainly be a case of adaptation, but both companies would be on the same hill and the copier would just be following in the path already taken by the other. If instead the products are seen as the hill climbers, then they are indeed climbing different hills, but copying the competitor would be the extinction of the design at the top of the lower hill while a new competitor appears simultaneously on the surface of the other hill. While fitness landscapes with attractor hills are useful metaphors, they require careful consideration about what exactly the evolvee is and what the landscape might look like so as not to get mixed up.

2.3.5 Coupled Fitness Landscape

In a system, all elements interact, but an adaptive system entails that those elements not only interact, but adapt to each other, reacting to selective pressures to become better suited to the system as a whole and the environment in which the system is situated. But every adaptation changes the environment and selection pressures acting on the rest of the system, so leads immediately to slightly different selection pressures. Since selection pressures are what shape the hills and valleys of fitness landscapes, the changes from every adaptation means the hills are alive, constantly jumping up or falling flat, moving around and changing shape to reflect the new selection pressures that are pushing towards the new attractor hill tops.

Everything is connected and nothing is stable. Referring to evolution as co-evolution emphasises that nothing exists or evolves in isolation. Every action of every element in an evolving system will have some effect on other elements. A comprehensive overview of the co-evolution literature is beyond the scope of this work,

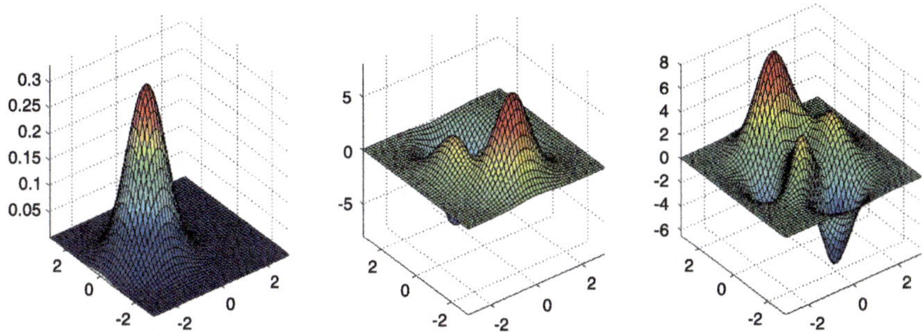

Fig. 2.2 Deformation of a fitness landscape

although it has already been thoroughly described in the literature of biology (Futuyma 1983; Jantzen 1980; Thompson 1994).

As nothing adapts in isolation, the fitness landscape can be recast as a coupled fitness landscape. This works particularly well in identified "arms races" where two species are each adapting to the last adaptation of the other and upping the selection pressure to adapt further, ratcheting up the height of the hill in each other's landscape. A classic example is that of the cheetah, which has adapted to run faster than any other land animal, and its usual prey, the gazelle, which cannot run as fast over the short distance but can maintain a very high speed over a much longer distance than the cheetah.

Hordijk and Kauffman (Kauffman and Johnsen 1991; Wilds et al. 2008) use the x and y axes of a coupled fitness landscape to represent the possible ranges of properties of two interacting species, such as the cheetah and gazelle. The z axis represents the combined fitness landscape of the two species, with peaks and valleys for combinations of different properties of the two species. The coupled fitness landscape is dynamic, with each adaptive step taken by one species distorting the fitness landscape of the other, and vice versa. This is illustrated in Fig. 2.2, where, going from left to right, the fitness landscape is deformed as species evolve and acquire new traits (e.g. higher top speed, more endurance, faster reflexes). If the gazelle adapts to run faster, the cheetah must adapt in some other way to deal with faster gazelles, reducing the gazelles fitness again. The responses and counter-responses can not be predicted in advance, as it might be higher speed, or better camouflage, or better sensory perception that provides the next temporary advantage. Of course, coupled fitness landscapes can be applied to socio-technical evolution as well, as companies battle to devise the next ingenious way to compete for customers, as new computer viruses develop sophisticated techniques to escape detection while virus detection companies also grow better at detecting, and as corporate loopholes are closed, only to allow some other loophole or tactic to be exploited.

But even coupled fitness landscapes are too simplified to deal with an entire adaptive system. They can only represent a few adaptations at a time, those assumed to be important, in a theoretically idealised population, in a way that can only look at the known past. Individuals, products, designs, companies, solutions or species, each with unique combinations of features and traits, can be represented on its own fitness landscape. When selection weeds out the losers, one evolvee may lose out, despite being the highest up the hill for a given trait, because of a fatal step down the hill for another trait. So every unique trait or feature could have its own fitness landscape, with all of them coupled to capture the fitness landscape of the evolvee, which in turn would be part of a larger fitness landscape that captures interactions between the evolvees. Just as the levels of nestedness in system boundaries can be understood to go on forever, so do the levels at which fitness landscapes can be coupled. It all gets incomprehensible very fast.

Irreversibility Attractors and fitness landscapes are theoretically related to another concept, called path dependency, also known as high switching costs (or sunk costs) (Economides 1996), group think (Janis 1982), and lock-in (Teisman 2005). Path dependency is captured by the idea that "history matters" (Buchanan 2000) because past decisions influence the future decisions to be made, which leads us directly into the concept of irreversibility.

Steps only go laterally or uphill, never downhill. But even a step laterally can not be reversed in an adaptive system because the landscape changes after every step. Whenever a move is made or an interaction is established, all other conceivable moves or interactions possible at the last step are no longer possible, although there is a whole new set of moves or interactions available. Any living or evolving process involves thermodynamically irreversible processes, so path dependency is "baked in" to reality at every level.[9] Thus any system that changes involves irreversible processes.

This irreversibility or path dependency applies to the system's overall behaviour, which can manifest itself in many ways. Physical systems lose mass or energy (Prigogine 1967), while social systems lose information. These losses also cause shifts in the landscape, affecting the future possibilities.

2.3.6 Intractability

Evolution is an algorithmic process (Dennet 1996) of variation, replication and selection across the evolutionary design space. Computational theory (Hartmanis et al. 1983) states that evolutionary problems are intractable,[10] that is, that future steps can *not* be calculated any faster than the time required to take that step. Intractability implies that the outcome of the evolutionary 'program' can only be found by completing its execution, which is to say, we just have to wait and see because predictions are impossible.

[9] This inherent reversibility is also called the "arrow of time", but if you start thinking that time arrows are baked into the fabric of reality you might be mixing your metaphors.

[10] Although normal English usage of "intractable" means uncooperative or stubborn, we are specifically using the computational complexity theory definition.

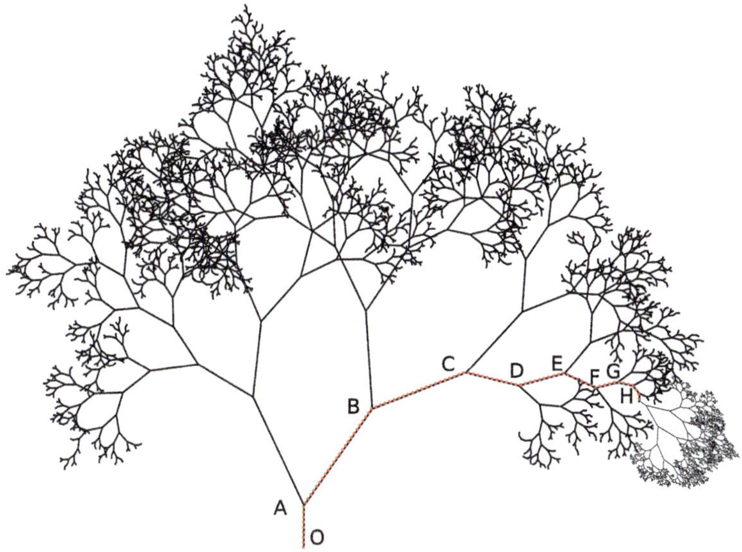

Fig. 2.3 An intractibility path

Thus, adaptive systems are impossible to predict with any reliability or exactitude, which means we face no small task when trying to understand and steer the evolution of adaptive systems, as discussed in Chap. 1. It can be mathematically proven that we will never truly know the precise effects of our actions, and we must thus act accordingly. An illustration of this process is presented in Fig. 2.3: Let us imagine a system being at some arbitrary point 0 in the system's history. At time A, something happens to move the system towards point A, forever excluding all the states toward which the system could have evolved but which are no longer possible. At time B, another interaction event happens, again excluding countless possible future states. As the system progresses in time, across points C, D, E, F, etc., more and more of the astronomically large number of possible system states are not able to come into being. Of course, at each time step, the same astronomical number of new possible states continuously becomes possible, as can be seen at point H.

We quickly realise that adaptive systems in nature are incredibly complex. Every component interacts with every other component in a possibly infinite number of ways (if not actually infinite, then certainly astronomically large and probably also Vast[11]) closing off some options forever and shifting the infinite possibilities of the next interaction in the future in mega-coupled fitness landscape. As if that were not enough, "the system as a whole causally influences the state of its own constituents, which in turn determine the causal powers of the whole system" (Kim 1999), which is philosophically quite problematic, despite seeming to occur quite regularly wher-

[11] Vast differentiates the super astronomically large from just ordinary large. For example, 10^{50} is a very, very large number. However, 1000^{1000} is Vast (Dennet 1996).

ever things are not run by philosophers who say self-causation is impossible. Some common examples include the mind-body link of psychosomatic disease like ulcers or hypertension, self regulation in social systems, the creation and endurance of norms and institutions, and, in fact, any adaptive system.

But before we jump ahead and look at how complex things behave, let's see more about what we mean by complexity.

2.4 Complexity

> "I shall not today attempt further to define the kinds of material I understand to be embraced within that shorthand description 'hard-core pornography'; and perhaps I could never succeed in intelligibly doing so. But I know it when I see it, and the motion picture involved in this case is not that."
>
> Stewart (1964)[12]

Complexity is, perhaps surprisingly, like pornography in that you can't really define it, but you know it when you see it. Definitions are observer-dependent and subjective, so rather than starting by trying to define what complexity is, we would first like to discuss what complexity is not, by looking closely at simplicity and complicatedness.

2.4.1 Simple

The most basic definition of complexity is that it is "not simple". Being the conceptual opposite of complexity, simplicity therefore requires a more careful examination, although it should come as no surprise that simplicity is also an elusive, slippery and uncooperative concept.

Nothing in the real world is ever really and truly "simple". Instead simplicity is relative, much like "big" or "wealthy", so a thing can only be judged as more or less simple than another thing. These comparative measures are necessarily observer-dependent, so the same thing can be judged as simple by one observer but as not simple by another, and at the same time they can both be complex. Already in the 13th century Dominican philosopher and theologian Thomas Aquinas argued in his influential work *Summa Theologica* that God is infinitely simple, but will look complex to the finite human mind, because every observer cannot help but see through the lens of unique perspective.

Setting aside the relative nature and the observer-dependence that this entails, simplicity can still be broken down into structural and functional simplicity. We can evaluate which of two structures that perform a given function is the simpler

[12]Concurring opinion in Jacobellis v. Ohio 378 U.S. 184 (1964) regarding possible obscenity in The Lovers.

structure. Likewise, for a given structure, we can compare which of two functions that it can perform is simpler. However, trying to determine the relative simplicity of two or more structures, each of which may or may not perform two or more functions, starts to lose meaning because simplicity in function and structure appear to be largely, if not totally, mutually exclusive. Thus, something both functionally and structurally simpler than another thing will be rarer than hen's teeth as any increase in simplicity for one measure will be cancelled out by the other. Thus, simplicity does not apply to a thing, but rather only to aspects of that thing in a particular context, which brings us back to observer-dependence and subjectivity. This may be useful to bear in mind when looking at complexity later in this chapter.

Functional Simplicity

> "Civilisation advances by extending the number of important operations which we can perform without thinking about them."
>
> Whitehead (1911)

Let's look at the light switch. They are easy to operate without much thought and creating bright, constant, useful light can be considered an important operation. The function of a light switch is straightforward and easy to explain, so light switches represent functional simplicity. But that functional simplicity requires quite a lot of non-simple structure, like a distant power source, a network of cables to transmit the power, various protections and fail safes to prevent dangerous surges or shorts, and an electrician to setup the switches in the house, among other things, in order to make turning on a light as simple as flicking a switch.

Before all of the structure of power grids and standard light switches was established, turning on a light involved a lot of steps, each of which would have been fairly easy and straightforward, but which required more work and a particular order of operation and which allowed opportunities for failure. Whitehead equates the advance of civilisation with the collapse of several functions, each requiring effort and attention, into one simple function by means of non-simple structures, mechanisms, procedures and tools to sequence, execute and monitor all of the sub-functions.

Simpler function can be very positive, worthy of motivating Whitehead's admiration, as when complex structure provides simpler function in medical devices (Burdulis et al. 2010). Manufacturing and industry have also benefited from the development of complex structure and complex processes which allow for impressive improvements in performance and waste reduction (Aldrich and Whetten 1981; Schonberger 1982). These benefits of simple function arise from bundling multiple necessary structures and functions together so that the relations between the required parts and steps can be more automatic, integrated and efficient, but simplicity of function is not always desirable. For example, Miller (1993) describes how over time, organisations and businesses increase the complexity of organisation and the simplicity of function to their own detriment by maintaining no longer necessary or successful structures. Further, when faced with complex situations and choices, people fall back on simple but suboptimal options (Iyengar and Kamenica 2010). The tendency to maintain complex, but no longer optimal, structures or to choose suboptimal solutions when faced with complex choices is related to the irreversibility and path dependencies created in adaptive systems. An adaptation that moves up a hill may be beneficial, but reaching the top of a hill always carries a certain risk of becoming trapped without any chance of further movement.

Structural Simplicity The simplest possible structure is a whole, without internal divisions or parts. Unfortunately, nothing known to man is truly whole and non-divisible. Atoms, derived from the Greek word $\alpha\tau o\mu o\zeta$, meaning indivisible, are a prime example of the lack of indivisibility, as they are now known to have internal parts which can be split, exchanged and recombined.[13] The closer we look at what we assume to be whole or structurally simple, the more internal divisions we find. It appears to be turtles all the way down,[14] although at the Planck length, we lose the ability to distinguish where one turtle ends and the next begins.

Cheerfully choosing to ignore the internal components, we can call the structure simple when it is reasonable to do so, which of course is entirely dependent on the observer and the context of observation. Water, for example, does have internal structure, but the arrangement of hydrogen and oxygen atoms is arguably irrelevant if we want to look at whether we need more or less water for growing tomatoes, so we can ignore the internal structure and treat water as simple.

However, the simpler the structure, the more likely it is to have many functions. Water, rocks, planes, wheels, and other simple structures are useful, even necessary, for loads of functions. Rocks, for example, can be projectiles, weights, tokens, building materials or pets, when googly eyes have been glued on, and many other things as well. These simple structures appear to have simple functions, but they are usually only one part of a complex structure to perform a larger function. For example, rocks are only useful as tokens in a complex system of symbolic representation including ideas of value, delayed reciprocity and the exchange of goods and services, while rocks as projectiles are not often just launched for their own sake, but as part of achieving military aims, hunting, games, sporting competitions or scientific experiments.

Thus, even simple structures, when repeated, linked or incorporated into sequences, quickly become complex. Classic examples include the repetition of simple DNA structures, which generate complex gene regulation functions (Tautz et al. 1986), and languages, where structural simplicity permits such complex functions as expressing everything ever said or written (Ferguson 1968).

Occam's Razor The law of parsimony, often known as *Occam's razor*, is often understood to advocate the simplest explanation when choosing from several competing hypotheses. However, this "simple" rule can make finding the simplest explanation very difficult. Whoever is holding the metaphorical razor must decide whether to favour the explanation with the simplest function or the explanation with

[13]And new divisions for those parts, and for the parts into which they are divided, are always lurking at the edge of theory.

[14]http://en.wikipedia.org/wiki/Turtles_all_the_way_down.

the simplest structure, which are unlikely to be the same explanation for any phenomenon of interest. In fact, most of the competing hypotheses will be almost impossible to compare for simplicity when both functional and structural simplicity are included, especially if the measure includes all of the associated and supporting structures and all of the alternative or possible functions.

Greenhouse Example A structurally simple greenhouse, essentially just a transparent enclosure, is not specific to any particular plant species, so has multi-functional potential and can be used to grow almost any kind of plant, or even multiple types of plants at once. There is even potential to add chickens or mushrooms for increased, complex function. But such a simple structure requires quite a lot of additional, albeit also simple structures, such as watering cans, harvesting tools, baskets or barrows for carrying the produce, and quite a lot of manual labour to operate. A modern high-tech tomato greenhouse on the other hand, which comes with specifically spaced racks, CO_2 transport tubes, temperature sensors, lighting, heating, aeration, pollination and watering systems, tomato harvesting robots and automated transport trucks with safety systems can only accommodate tomatoes.[15] The increasingly complex structure means that it has the simpler, more specific function of "growing tomatoes", which it can do extremely efficiently and effectively, but requires far more complex internal structure to achieve it.

As a second example, let's compare an average smartphone vs Casio Model AS-C. The Casio AS-C only calculates a few limited mathematical operations, while the smartphone has many, very diverse functions including a calculator. From the perspective of a single function, perhaps calculating the amount everyone must pay toward a shared bill at a restaurant, they are equivalent in functional simplicity. They can both divide the total cost of the bill by the number of people at the table, so the smartphone will look far more structurally complex, even annoyingly so as you might have to go through several menus to get to the calculator function while the Casio AS-C need only be plugged in. However, if all of the functions available on a smartphone are considered, then it appears far simpler to carry one small device in place of carrying a series of simple but separate devices such as a telephone, calculator, camera, mp3 player, laptop, etc.

2.4.2 *Complicated*

Before we go on to discuss what complexity is, we want to first introduce a special type of complexity, know as complicated. Complex and complicated are both non-simple, but the important distinction between them is not an inherent quality, but is

[15] Modern greenhouses can of course be adapted to grow other crops. Some of the structures and systems would become completely redundant, others would need to be added, and still others would require some adjustments. Like the structurally simple greenhouse, there is potential for more, but the path dependency means that it will be much more costly to switch to a new crop if there is a risk that high cost investments, like a tomato picking robot, will become utterly useless.

a matter of process, change and experience. George Whitesides[16] argues that simple things are:

- reliable, predictable;
- cheap (money, energy, etc.);
- high performance or value/cost; and
- stackable, able to form building blocks.

His view of simple does not distinguish between structural or functional simplicity, instead highlighting elements of each. Reliable and predictable, for example, apply far more to simple structures[17] than to complex structures, but, like the light switch, even complex structures can become predictable, cheap, high performers which can be used as elements in larger structures.

As Whitehead remarked earlier, the hallmark of civilisation is the change in effort required to do something important. This move is not from the complex to the simple but from the complicated, or both structurally *and* functionally complex, to the simple, where either the structure or the function is judged as simple. Many new, cutting edge, experimental or unfamiliar socio-technical innovations are not predictable or reliable, not cheap to make or use, not high performing and not useful as part of larger systems. These systems, made of many parts, involving long sequences of actions, demanding training, and requiring a high level of vigilance to maintain or control, can not be judged as structurally or functionally simple, and so are complicated. Complicated system are argued to be more difficult to understand than complex systems (Allen et al. 1999), possibly because, as brand new entrants to a fitness landscape, the direction of motion and the effects of any steps taken are uncertain.

What counts as a complicated system is highly observer-dependent.[18] Cars and airplanes are generally considered to be good examples, but others might be the constantly changing rules in Formula One racing, the unwritten and ever shifting norms of fashion, etiquette, and cool music, the fluctuating and labyrinthine financial regulations, or the steady stream of upgrades to high tech software programs. These examples are clearly not structurally simple as they have thousands of rules, elements, subsystems and interacting parts, the removal, misuse or malfunction of any of which could result in a non-operational vehicle, the imposition of a ten second penalty, a social faux pas, astounding financial losses (or gains, if you can spot the loopholes), or surprising software bugs. But in addition to being structurally complex, they are also functionally complex because they require significant and unending effort, attention, training, vigilance and maintenance. Driving a car or taking advantage of the best corporate tax schemes is just not as easy or effortless as flicking a light switch.

[16]http://www.ted.com/talks/george_whitesides_toward_a_science_of_simplicity.html.

[17]How often do really simple structures fail? A rock, for example, tends to be flung from a catapult with a high degree of reliability.

[18]What else did you expect?

The epistatic relations between the many structural elements of complicated things mean that the parts, even if optimised individually, may perform in unpredictable and sub-optimal ways when put together, so any new design will be complicated. But when complicated systems, as a whole, are used over a long period of time, subject to relatively constant selection pressures in steady conditions, they become seamless, effortless and almost invisible as a part of the background of everyday life. Engineers or other experts often idealise either the functions or the structure of complicated systems as relatively simple, downplaying the importance of variations, interactions, or the effect of the environment to see the system as more isolated, mechanistic, and with out any surprising behaviours or flaws. While these idealisations are useful in the design process, to the unfamiliar and non-engineers, the complicated systems remain bafflingly complex in every way. Yet as a design matures, it is tested in many new ways, is subtly refined, grows more prevalent, reliable and inexpensive, performs better in relation to the value, and can serve as a component in larger, newer and more complicated systems. Structural elements may be removed if they can be reliably expected to exist in the environment, simplifying the structure, or the structure may adapt to be more complex, but better integrated, so that the function is simpler. Even if the structure or function does not actually change much, the familiarity of continued use will affect the observer-dependent judgements so that either the many parts come to be viewed as a simple unit or the functions come to be viewed as straightforward and easy.

Greenhouse Example Automation is at the leading edge of an increase in greenhouse complication. Automatic tomato and flower picking robots can not only pick the produce, but also place it on automated conveyor belts which feed into automated packaging machines. These systems have many specific parts, such as conveyors, chutes, clamps, sensors, motors and robotic arms. When working correctly, operation is very efficient, but there has simply not been enough time yet to test every possible situation or combination of factors. Consequently, not all of the "bugs" have been worked out and the failure of something as simple as a ball bearing in one part of one subsystem has the potential to disrupt the entire operation.

However, automated watering systems used to be cutting edge, full of bugs and with enormous potential for risk, but are now seen as commonplace. The continued use has resulted in predictable, inexpensive and highly valuable performance, so they are no longer complicated, but functionally simple parts of a larger complexity.

2.4.3 Complex

Finally, we come to discuss complexity. We already know quite a bit about complexity because real world systems are inherently complex, despite the tendency to idealise them as isolated, mechanistic and fully knowable. Adaptive systems are even more complex because the relationships between elements are so pervasive, subtle, and impermanent. Change is inevitable, and every change rewrites the rules of the game in some small way. We also know that complex is the opposite of simple, but that new, complicated additions to the system adapt themselves and force the adaptation of their environment until they are so embedded that they look simple when viewed in the right way. And finally, we know that viewing in the right way is key to seeing a whole made of parts or a part in a larger whole, as complex or simple, or as caused by rules or as the cause of those rules. The importance of views makes complexity so infuriating impossible to define because, as Mikulecky (2001) states:

> "Complexity is the property of a real world system that is manifest in the inability of any one formalism being adequate to capture all its properties. It requires that we find distinctly different ways of interacting with systems. Distinctly different in the sense that when we make successful models, the formal systems needed to describe each distinct aspect are *not* derivable from each other."

Formalisms, or formal systems of capturing statements, consequences and rules, that are not derivable from each other, such as mathematics and psychology, capture different truths about a system. To really describe a complex system more than one formalism, incompatible as they are, must be employed because only the multiple viewpoints of different formalisms can come close to seeing the system as a whole and a part, complex and simple, the cause and the effect, *at the same time*. Checkland goes even further to say that "human activity systems can never be described (or 'modelled') in a single account which will be either generally acceptable or sufficient" (Checkland and Checkland 1999, p. 191). Knowledge from various domains and disciplines must be integrated to begin to describe the properties and behaviour of a system in a more adequate, acceptable and sufficient way. As most people master a limited number of disciplines and formalisms, the increased attention to complex systems will demand an increase in interdisciplinary cooperation, although every account will still face criticism, probably from those whose formalisms or models were not included, as insufficient, which must be balanced against the increase in complexity from the inclusion of additional formalisms.

Dynamics One important truth of complexity is that it happens in many dimensions at the same time, and one often overlooked dimension is time. Many attempts have been made to understand why we have the complexity we see in the world, especially as complexity involves so much apparent simplicity due to the balances between complex structures and complex functions over time.

Smith and Szathmáry (1997) argued that important transitions in how information is transmitted represent the key points in the development of complexity, typically when simple structures with multiple functions developed more structural complexity by apportioning each new structure with fewer, simpler functions. Allen et al. (1999) on the other hand, suggests that simple structures with simple functions multiply and compound until reaching a critical point after which the parts restructure into a hierarchy. They consider the addition of new structures at any level to be an increase in complicatedness but that the increase in levels, or the deepening of the hierarchy, to be an increase in complexity. Although they use complicated in

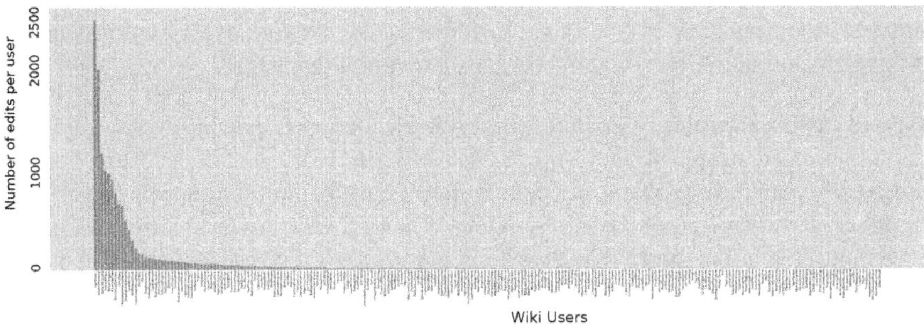

Fig. 2.4 Power law observed in the edit frequency per user on wiki.tudelft.nl

a different way than we have so far, their use also supports the idea that new additions to a system are complicated but that over time the system adjusts and this complicatedness disappears into the total system complexity.

Self-similarity or Scale Invariance Fractals are non-Euclidean, irregular, geometric structures where each individual part is, at least approximately, a reduced-size copy of the whole fractal. This recursive self-similarity is true of complex systems because they are nested, with each level being the lower level of a larger system, or the higher level comprised of smaller systems. But not only are complex systems self-similar, or scale invariant, in structure, but also in behaviour, so that the same patterns, shapes and proportions hold true of the output of the system, no matter the scope of the perspective. An example that you might find outside your front door would be the formation and propagation of cracks and tears in concrete slabs which follow power law proportions, also known as the Pareto distribution, the 80/20 rule or long/fat tail. The frequency of the cracks varies as a power of an attribute, such as the size of the cracks. Thus, very large cracks are relatively rare, numerically overwhelmed by the small cracks, yet the big cracks, rare as they are, overwhelm the total size of all the small cracks added together. This relationship is true if you look at the cracks in just one square meter or at all of the cracks in the entire street. Given the ubiquity of complex systems, it should come as no surprise that these scale invariant relationships are observed in a wide range of phenomena, from craters on the moon to the distribution of wealth in an economy and edits in wikis, see Fig. 2.4.

But scale invariance works in the dimension of time as well as space, so that the relationships between the frequency of an event and some attribute of that event hold the same relationships at any time scale. For example, avalanches occur at any size, from the catastrophic collapse of entire hillsides to the tiny movement of small clumps of earth or snow. The likelihood of an avalanche is in power law proportion to the size of the avalanche, so the relationship between the frequency and size of avalanches observed is the same regardless of whether you look at data for a year, ten years or 10,000 years. Other examples of scale invariance in time include the frequency and duration of network outages on the Internet, the frequency and number of journal articles citations, considered in the network of all citations among all papers, and the distribution of word use in natural languages.

Greenhouse Example Greenhouses display scale invariance in several ways. As greenhouses are nested systems, they reveal self-similarity at various levels. For example, a greenhouse growers association might have committees devoted to certain topics or aspects of greenhouse operation. When moving down to large individual greenhouses that belong to that association, you might find an individual manager devoted each of those same topics, replicating the structure of the next level up. In smaller greenhouse, there might not be managers dedicated to one topic or aspect of operations, but there will be some replication of the structure, even if it is only that paperwork is clustered into file folders in roughly the same divisions as the committees at the association level.

But greenhouses also display scale invariance in behaviour, space and time. Greenhouse owners, like all people, show power law distributions in the number of contacts they maintain and the frequency of use of those contacts, so that a very few greenhouse growers have large networks of contacts, while most greenhouse growers have far fewer contacts, and that each grower uses some of their contacts very frequently, and the remaining contacts only very infrequently. Further, these relationships are held if you look at any scale, from the local and present to the national and historical. Many aspects of greenhouse operation or behaviour, from the size of greenhouses to the energy use of the devices inside the greenhouses, reveal the complexity of the greenhouse systems by displaying scale invariance and self-similarity.

2.5 Complex Adaptive Systems

Putting the three previous sections together gives us complex adaptive systems, which John H. Holland (Waldorp 1992) defines as:

> "[...] a dynamic network of many agents (which may represent cells, species, individuals, firms, nations) acting in parallel, constantly acting and reacting to what the other agents are doing. The control of a complex adaptive systems tends to be highly dispersed and decentralised. If there is to be any coherent behaviour in the system, it has to arise from competition and cooperation among the agents themselves. The overall behaviour of the system is the result of a huge number of decisions made every moment by many individual agents."

This interest in and acceptance of complexity heralds the rise of a new paradigm, and researchers are aware that there is something important going on, although there is not yet a consensus as to what exactly it is. As a new scientific paradigm, complex adaptive systems is a lens for looking at the world and the way it operates that allows, even requires, a multitude of perspectives and formalisms. No single approach or description will be adequate to capture the richness of complex adaptive systems interactions at many dimensions and across several levels, creating dynamic emergent patterns from local interactions between system components (Holland 1996; Kauffman and Johnsen 1991; Newman 2003).

Although we have already been treating greenhouses as examples of complex adaptive systems throughout the chapter, they are also socio-technical systems with both physical and social co-evolution. As complex adaptive systems is displacing older paradigms that equated understanding with simplification, explanation with a single description, and strict isolation between physical and social elements of systems, engineers need to be aware, but not overwhelmed or discouraged by the links between the physical and social. This means that the design of artifacts now faces new dilemmas, as technologies are seen to be influenced by as well as influence the people who interact with or use the technology.

Greenhouse Example Greenhouses are uncomfortably hot places, and the work that needs done nevertheless requires a high level of physical exertion. As a result of incalculable factors, this is currently perceived as undesirable work, reducing the supply of readily available labour. As a consequence, automated or robotic systems are increasingly attractive to greenhouse growers as they do not feel stigmatised by societies disdain for sweaty conditions or heavy lifting. But the use of tomato picking robots, automated packaging machines and self-guiding product carriers reduces, but has not yet eliminated, the need for human employees. The fewer jobs remaining are now more monotonous, lower skilled, and more isolated, further suppressing demand and wages for greenhouse jobs and reinforcing the perception that these are unappreciated, difficult and low paid jobs.

Did the designers of automated greenhouse systems consider the effect their innovations might have on the social elements of the socio-technical system? Would the results be different if they had? Or are attempts to directly influence the non-physical aspects impossible or unethical anyway?

Of course, co-evolution is intractable. We will never know what would have happened if society had instead perceived a hot, sweaty, physically demanding greenhouse job as a noble, enviable and rewarding position. Robots might be far less attractive if young and strong people competed for the chance to spend their days lifting weight and sweating out skin impurities while being paid to produce food for the benefit of all of society. But again, unpredictable as complex adaptive systems are, engineers in this alternate reality would be busy devising some other tools that influenced wages, labour or other social aspects of the system, with perhaps no net difference.

2.5.1 Chaos and Randomness

One of the basic mechanisms at play in all complex adaptive systems, and one of the reasons we will never know if engineers in that alternate reality could really have a net effect on greenhouse horticulture as a sector, is chaos. While not all chaotic systems are complex adaptive systems, all complex adaptive systems contain chaotic elements. Chaos is a large field of study, that we will not attempt to cover exhaustively here. Instead we will highlight and discuss the main points relevant

Fig. 2.5 The Mandelbrot set fractal

to socio-technical systems. For more background please refer to Gleick (1997) or Kellert (1993).

Chaos can be defined as complex behaviour,[19] arising in deterministic, non-linear dynamic systems, when relatively simple processes or rules are repeatedly applied. Chaotic systems display, among others, two special properties:

- sensitive dependencies in initial conditions
- characteristic structures

Repetition Chaos arises from the repetition, iteration or recursion of simple rules, formulae, processes, or mathematical functions, such as logarithmic maps or fractals. For example, repeatedly reevaluating the complex function $Z = Z^2 + Ci$ gives rise to the Mandelbrot (Mandelbrot 1983) fractal, as seen in Fig. 2.5. The iterations of the simple processes of selection, replication and variation, allow for chaos to develop in adaptive systems, driving some of the complexity of complex adaptive systems.

Deterministic The state of any dynamic system changes over time, according to some rule or procedure. If the changes have no trace of randomness and are instead completely controlled by the rules or procedures, then the system is deterministic, meaning that a given cause always has a clear and repeatable effect. Fractals, both deterministic and chaotic, are not in any way random. In fact, randomness is far harder to produce than it seems as randomness is completely without cause and contains absolutely zero information. No known model is capable of producing true randomness, and your computer cannot produce an authentically random number.[20] The only suspected source of true randomness in the universe is the decay of radioactive atoms driven by quantum fluctuations (Green 1981).

Initial Conditions But if chaos is deterministic and randomness is not the source of complexity in complex adaptive systems, why are they intractable and unpredictable? While being non-random and fully deterministic, the iteration of rules on

[19]Chaos is a complex behaviour, but chaos is not the only mechanism driving the complexity of complex adaptive systems, nor is chaos the same as complexity.

[20]Get your truly random numbers here: http://www.fourmilab.ch/hotbits.

the system magnifies the minute differences between two starting conditions, potentially leading to very different outcomes and the appearance of unpredictability. Often referred to as the *butterfly effect*, this sensitivity to initial conditions means that seemingly insignificant differences, such as the rounding off of numbers in calculations, tiny errors in measurements, or a change in what appears to be a totally unrelated factor, can be sufficient to set the system into a different state, making them appear to act randomly and without reason. With only finite information on the starting conditions, the exact state of a chaotic system cannot be predicted with any certainty, and the uncertainty grows with the distance of the forecast. This is why specific weather predictions beyond a week are no better than guesses.

Attractors Characteristic structures, the second property of chaotic systems above, means that chaotic systems tend to converge towards certain points or regions in the systems state space, over time, called attractors. Thus, while sensitivity to initial conditions means that the exact states cannot be predicted, attractors mean that some very large sets of initial conditions converge on a single chaotic region and this convergence can be predicted with some reliability. Usually, system outputs contain multiple attractors, and some contain repellers as well, which the system seems to be unable to approach. Dynamic systems will have attractors that shift over time, displaying varying intensities and duration of attraction, all influenced by the complexity and adaptations of the system.

These are the same attractors that form the hilltops in fitness landscapes. In adaptive systems, the attractors are a consequence of selection pressures that act on differences in fitness, but non-adaptive systems have attractors too if the rules of the system interact in such a way as to create one or more basins of attraction.

Instability and Robustness When a chaotic system suddenly changes from one attractor to another with only minimal parameter changes, it is called instability. There are many examples of instability, where an apparently consistent and predictable system appears to suddenly change gears and move with a unsettling inescapableness along what previously appeared to be an unlikely path. The human heartbeat, for example, displays a change from one attractor to another when it rapidly increases after hearing a sudden noise. Large crowds are notoriously unstable, suddenly erupting into riot, moving in new directions, and forming crushes or stampedes under certain conditions that do not appear to be very different from the conditions in which crowds of the same size act peaceably. In structural engineering, a structure can become unstable when an applied load crosses a threshold and the structural deflections magnify stresses, which in turn increases the deflections.

Instability can be seen as the opposite of robustness. Note that robustness is not the same as stability, as systems can be simultaneously robust and not stable.[21] A system is robust when it is close to or at an attractor because very few parameter changes can cause a deviation from the path to the attractor (Callaway et al. 2000).

[21]The terms stability and instability seem like each others opposites, when in fact instability is opposite robustness.

However, robustness is a not a general concept, as large changes in some parameters cannot make the system deviate from its path to an attractor, while only very slight changes in another parameter might cause the system to change to another attractor entirely. Robustness is a measure of how the system performs under stress, when confronted by extreme inputs or shocks from the environment, only for particular variables. The Internet, designed to function even if large parts of it are destroyed, is robust against physical attacks, power outages and disruptions, but is weak against sudden rushes of net traffic which can spread the disruption from overloaded sites. Body temperature is robust against changes in temperature of the environment, but can be disrupted by illness, anxiety or even stress. A less positive example is economic lock-in, when a customer is so dependent on a supplier for products and services that the switching costs of moving to another supplier outweigh the benefits.

The combination of chaos, instability and robustness are important concepts for complex adaptive systems and socio-technical systems. Importantly, robustness and instability need to be seen in relation to specific parameters, and to be viewed in context, as the ability to change suddenly or to resist large changes can both be seen as good and bad. Any attempt to engineer, shape or steer a complex adaptive systems must be careful to analyze exactly which parameters the system is robust or instable in relation to, as large intentional changes in the wrong parameter can have little or no effect, while small, accidental and unintended ones can dramatically affect the system.

Greenhouse Example Greenhouses in the Westland often use combined heat and power units to produce heat, electricity and CO_2. At certain times of the day, they produce more electricity than they need, so sell this extra power back to the regional electricity grids, making the power generation capabilities of the region distributed and therefore robust against a catastrophic loss of power. On the other hand, sometimes the greenhouses need electricity, but cannot help producing superfluous heat and CO_2 as well. This wasted production contributes to the total greenhouse gas emissions of the region, which appear to be very resistant to all efforts at reduction, indicating that a very strong attractor makes the system robust to changes that would result in a net decrease in CO_2.

Instability in the greenhouse horticulture sector is readily apparent in the prices paid for products, which can change suddenly and drastically. While the prices paid for the goods can change suddenly, switching from one attractor that drives prices up to another that drives them down, the invested time, effort and money that has already gone into producing the flowers or vegetable is far less unstable, so a mismatch between production costs and selling prices is always a risk. Unfortunately, the limited shelf life of the products means greenhouse farmers are unable to wait for prices to improve.

2.5.2 Emergence, Self-organisation and Patterns

Emergent behaviour or emergent properties are overall system behaviour of a complex adaptive systems. Emergent behaviours contain *no magic* because they are only the motion toward attractors, or away from repellers, although they are rarely obvious or predictable. Instead, the apparently magic new characteristics or phenomena are only the logical consequences that become apparent once the organisational structure and interactions of the system are constituted (Crutchfield 1994; Morin 1999). These phenomena *cannot be deconstructed solely in terms of the behaviour of the individual agents* (Jennings 2000) and *would not arise if isolated from the organising whole* (Morin 1999). Indeed, the emergent properties of systems are lost when the system is broken down into parts and parts removed from the system lose the emergent properties they previously possessed.[22] Although an emergent property cannot be found in any of the component parts, nevertheless, emergent properties can appear or disappear with the gain or loss of a single element, depending on where the emergent behaviour is in relation to the various attractors of the system, as chaotic systems are always instable or robust to particular parameters. Examples of familiar emergent behaviours include traffic jams, for which it makes little sense to examine the actions of individual cars, schooling, swarming or flocking behaviours in social animals, which generally have no centralised control and yet behave cohesively as a group, or stock markets, which aggregate the actions of many traders, all of whom have limited knowledge and operate under regulations, yet lead to wildly different results from one day to the next.

Emergent behaviour tends to be easier to recognise or simpler to understand—and potentially more insightful—than the collection of processes that cause it, leading many emergent phenomena to remain as "black boxes" to observers. Human consciousness, for example, is argued to be an emergent behaviour (Dennet 1996) and cannot be understood in terms of the individual parts of the brain. Institutions, governments or corporate boards display unpredictable, emergent system output when establishing policies because "a decision is an outcome or an interpretation of several relatively interdependent streams within an organisation" (Cohen et al. 1972), each with incomplete information and unclear priorities (Lindblom et al. 1980). Economic literature also suggests that externalities are undesired emergent properties, although not all externalities are entirely negative, as when neighbourhood house prices are increased through the many, distributed actions of dedicated home gardeners. Emergent properties are what we look for when studying socio-technical systems and the evolution of these systems. Although perhaps not often framed as such, most of the decisions we take in life, and certainly the decisions made by authorities, are geared toward bringing about or enhancing desired emergent properties, like sustainability, while preventing the undesired ones, such as pollution.

Greenhouse Example The selling price of a given greenhouse horticultural product is emergent, and depends on the interaction between costs of growing substrate, greenhouse technologies, labour, and energy, but also on the behaviour of other tomato growers, the demands and leverage of supermarkets or other retailers, the

[22] Consider, for example, how the emergent property of living is lost if an organism is dissected, and how an amputated part ceases to be alive after removal.

shopping and consumption patterns of consumers, the implementation of legislation, the effect of foreign competition, new food fads and many other things. The prices are not determined by any centralised agent, nor can the final price be attributed directly to any one cause or action. Were the current system to be dismantled and all the parts isolated, for example if no communication or delivery of product were allowed between growers, retailers and consumers, not only would the prices not be set as an emergent property, but the entire concept of price would have no meaning. The participants in the system change all the time, as greenhouse companies start up, close, change hands or switch to new business models, as do the retailers, the consumers and the legislators, so the system is robust to the loss of any single participant. Yet it could all collapse entirely, or change beyond recognition, if some element, like the capacity for refrigeration were to be removed, or some as yet unknown new element were to be added that gave fresh products an unlimited shelf life. As it is, the interactions that determine the selling prices are contingent on the expected shelf life of the products and a sudden change in that would rewrite the rules entirely.

Self-organisation A particularly important and interesting form of emergent behaviour, is self-organisation, the process by which a system develops a structure or pattern without the imposition of structure from a central or outside authority, or when a system displays a different output as a result of internal processes (Prigogine and Stengers 1984; Kay 2002). For example, in morphogenesis, an embryo develops toward a fully functional organism by self-assembling from a single fertilised cell (Campbell 2002) while autopoiesis means that societies develop and impose limits to individual choice, which provides a more predictable, self-steering system (Luhmann 1995).

Structure and organisation can be very beneficial, durable and self-reinforcing, so self-organisation can be an adaptive response. But self-organisation also occurs in systems that are not generally considered to be adaptive, such as crystal growth, galaxy formation, micelles and cellular automata. Thus, self-organisation is not adaptive on its own, but is potentially adaptive, depending on the environment and the current state of the system. Influencing socio-technical systems and the evolution of these systems seeks to match up self-organisation and adaptation as much as possible. Self-organising behaviours that are also adaptive to the pressures we want apply to a system are essentially "for free", allowing relatively small modification of system components and their interactions to achieve a great degree of the desired organisation and regularity in the system.

Patterns Patterns are merely something we observe as standing out in contrast with background "noise". Living organisms have evolved the ability to detect regularity, even to the point that we are sometimes unable to *not* see a pattern in coincidences (think about conspiracy theorists), to ignore a change from one pattern to another (as when we can't sleep in a new place because we miss the sounds of traffic that we are used to), or to avoid tapping our foot when a catchy tune comes on. The regularity in a pattern makes it possible to compress it for more efficient understanding, storage or transmission. The pattern also allows for better than chance predictions about what comes next, which is impossible if there is no regularity in the data.

Computers, on the other hand, do not see patterns unless told to look for them, instead taking all the input as a whole, which is more time consuming, but less likely to produce faulty analysis. However, life being limited by time, evolution favours the efficient storage and transfer of information (DNA and natural languages are both good examples), faster response times (better not to wait until you see a predator if you could recognise the pattern of footsteps behind you instead) and the ability to develop predictions or hypotheses ("food has been found here this time last year, so could be found here again now"), even at the expense of potential loss of accuracy or the risk of false pattern detection. As patterns involve repetitions, they appear quite often in nature and evolving systems, and the detection of these repeated patterns also occurs regularly.

But these patterns do not just appear in nature, they emerge. That is, the regularity of interactions in the system leads to repetition of properties, behaviours, and structures, which are all detected as emergent patterns or as organisation in a dynamic system. The different system levels are apparent to us, not because they are true and real distinguishable levels, but because we see regularities in the interactions, over time or space, that lead us to observe a pattern. Thus, any pattern detected is highly observer-dependent and would not be seen, or would be seen slightly differently, by another observer with different access to the information or a distinct world view. As the system continues to evolve, the system levels may appear to break apart, grow, shrink or add additional levels as novel patterns of interaction emerge. And as the patterns emerge, they can serve as the input for the emergence of new patterns because the system itself capitalises on any regularities present, amplifying the patterns across different system levels.

2.6 Modelling Complex Adaptive Systems

Knowing all of this about systems, adaptations and complexity is great, but how can we use this knowledge? One proposal is that through modelling these systems in light of the principles of complex adaptive systems, we can better understand the specific systems and how to interact with them in order to achieve goals. In this section we will first discuss aspects of modelling complex adaptive systems in general, before moving to the theory behind agent-based modelling. Practical aspects of the creation of agent-based models are described in Chap. 3 in great detail.

2.6.1 What Does a Model of a Complex Adaptive System Need?

Models are the formalisation of a modeller's interpretation of reality, and so are not one, but *two* steps removed form the real world. The challenge therefore is to carefully balance two important, yet conflicting, needs when modelling a complex adaptive systems. The model must be complex enough to represent the system as well as possible and it must be as simple as possible in order to facilitate a greater understanding or the ability to change the system.

The first requirement is formally expressed by Ashby's Law of Requisite Variety (Ashby 1968), a commonly used formulation of which states:[23]

> "a model system can only model something to the extent that it has sufficient internal variety to represent it."

Thus, to be a successful model *of* a complex adaptive systems, the model must also *be* a complex adaptive systems. This need for accuracy and complexity in the model is directly at odds with the other need, which is to be simple enough to give insight into the system that is not attainable through observation of the system itself. Thus, the model must simplify reality in order to be useful, but not so much so that it is useless. Just as a map is a useful simplification and miniaturisation of a place, a model must be gainfully simplified by, among other things, defining system boundaries, level of observance and context.

> "All models are wrong, some are useful ..."
>
> Box (1979)

As already discussed above, each model is two-fold simplification of the reality, and, as a consequence, is wrong. However, even when wrong, a model can still be useful if the simplifications are only where appropriate to achieve the task at hand. A model will not be useful if it ignores crucial aspects of the real world system just because they are difficult or ideologically unpalatable. Likewise, replicating too much detail at a very low level, or refusing to include essential details from lower levels are examples of how greedy reductionism or extreme holism are unhelpful oversimplifications and can undermine the system representation. Therefore, we must find that tricky balance between the accuracy needed to reproduce complexity and the simplicity needed to gain any novel insight by following some clever advice. It has been said that the usefulness of a model can be estimated by the speed by which it is replaced. The models that provide the most insight and teach us the most tend to be replaced the fastest as a consequence of their high utility. Thus, we should follow the advice of a clever man when building models:

> "Everything should be made as simple as possible, but no simpler."
>
> Attributed to A. Einstein

In order to do so, we have found that every complex adaptive systems model should contain the following three main properties:

Multi-domain and Multi-disciplinary Knowledge Although any particular model can only be considered a single formalism, they can be used in multi-domain and multi-disciplinary ways to capture multiple formalisms. For example,

[23] http://pespmc1.vub.ac.be/REQVAR.HTML.

they can be developed with insight from multiple viewpoints, fields of study or experts with varied interests. They can also be used in context with work founded in non-derivable formalisms for a balanced approach to the inherent complexity of the topic, or as one part of a series of models, with each one incorporating new aspects that make it a slightly different formalisation.

Generative and Bottom up Capacity The central principle of generative science is that phenomena can be described in terms of interconnected networks of (relatively) simple units, and that finite, deterministic rules and parameters interact to generate complex behaviour. Most of generative science relies on the idea that "If you did not grow it, you did not explain it!" (Epstein 1999) and thus seeks to "grow" a given macroscopic regularity from an initial population of autonomous agents or to explore the range of behaviours a well understood, well described population of agents is capable of under different conditions. While not every behaviour that can be grown is necessarily also explained, the generative science approach means that if done well and founded on a rich theory of complexity and complex adaptive systems, modellers can attempt to build understanding from the bottom up.

Adaptivity The model must also be adaptive, with a capacity to evolve over time. There must be selective pressures to respond to, and a way to introduce variations for the pressures to act on. Ideally, the selective pressures should also be capable of shifting in response to the changes in fitness, making the entire system adaptive, although this is far more difficult to analyze and interpret and may push the balance of the model toward capturing the complexity of the system at the expense of insight that might be gained by simplifying the adaptivity.

Modelling Options Reviewing the many modelling techniques available would be outside the scope of this work. Some, like statistical thermodynamics, and pattern recognition tools such as neural networks, are unsuitable for modelling complex adaptive systems, as they are clearly not generative. Others, such as computable general equilibrium (Jones 1965; Leontief 1998), dynamic systems (Rosenberg and Karnopp 1983; Strogatz and Henry 2000), and system dynamics (Forrester 1958; Forrester and Wright 1961), are based on mathematical models based on an top-down paradigm and on a assumption of static system structure. Discrete event simulation (Boer et al. 2002; Boyson et al. 2003; Corsi et al. 2006; Gordon 1978) comes closer, being capable of both generative and dynamic system behaviour, but fails to suit our needs as the entities are very passive representations and cannot quite capture some of the necessary decision making. Agent-based modelling (Jennings 2000; Rohilla Shalizi 2006), however, has everything we need with its explicitly bottom-up perspective. The individual agents, whose algorithmic nature allows many different formalisms, act and react according to internal rules to produce the over all emergent system behaviour.

2.6.2 Agent-Based Modelling

Of the presented tools, agent-based modelling is the most suitable for modelling a complex adaptive systems because it is the only one that satisfies Ashby's requirement. In the words of Borshchev and Filippov (2004):

> The "agent-based approach is more general and powerful[24] because it enables the capture of more complex structures and dynamics. The other important advantage is that it provides for construction of models in the absence of the knowledge about the global interdependencies: you may know nothing or very little about how things affect each other at the aggregate level, or what the global sequence of operations is, etc., but if you have some perception of how the individual participants of the process behave, you can construct the agent-based model and then obtain the global behaviour".

Before we get into the nitty gritty of exactly what agent-based modelling is and does, we should explore a bit of its past. The first inklings of distributed computation that later came to underpin agent-based modelling appeared in the 1940s when John von Neumann conceptualised the Von Neumann machine (Von Neumann and Burks 1966), a theoretical device capable of self-replication using raw materials from the environment.[25] The notion was further refined by Ulam with the creation of the computer implementation which he called *cellular automata* (Burks 1970).

Constraints on computer power at the time meant that cellular automata remained as mere mathematical curiosities until Conway published his "game of life" (Conway 1970), a 2D cellular automata. The game of life demonstrated the extremely broad spectrum of behaviour that could arise from very simple rules. Thomas Schelling's segregation model (Schelling 1971) further advanced the possibilities. Although initially, played on a paper grid with coins, the segregation model showed some aspects of complex adaptive systems which were more clearly realised after it was later transformed into an agent-based model.

These early frontrunners paved the way for diverse and numerous explorations of the possibilities as computational power grew throughout the 1980s, including Robert Axelrod's prisoners dilemma model (Axelrod 1980) and Craig Reynolds' Boids (Reynolds 1987), a bird flocking simulation. The 1990s saw the spread of such models coinciding with an increase in the ready availability of vast amounts of computational power, while tools like Swarm, NetLogo and Repast lowering the programming barrier, allowing the agent-based modelling field to grow explosively. Currently, the field can boast a number of dedicated journals and with a series of high level articles appearing in journals such as Nature (Buchanan 2009; Farmer and Foley 2009) and The Economist (Economist 2010).[26]

---

[24]than System Dynamics, Dynamic Systems or Discrete Event Simulation.

[25]It is interesting to note that Van Neumann's idea is close to becoming a reality some 60 years later with the rise of open source 3D printers, which are currently able to build 90 % of themselves.

[26]Perhaps the economic crisis of 2008 could have been avoided or minimised if a complex adaptive systems approach was more widely used or appreciated. Although, we will never know, irreversibility being what it is.

Computer power is now ramping up and the theories of systems, complexity and generative science are gaining traction in science. Although previously too difficult to reason about, much less model, agent-based models provides a new tool for the exploration of these topics. The focus is on the interactions of the agents, which Stuart Kauffman says is "a thing which does things to things" (Rohilla Shalizi 2006). Furthermore, Rohilla Shalizi (2006) states that:

> "An agent is a persistent thing which has some state we find worth representing, and which interacts with other agents, mutually modifying each other's states. The components of an agent-based model are a collection of agents and their states, the rules governing the interactions of the agents and the environment within which they live."

Another perspective is provided by Tesfatsion (2007):

> "In the real world, all calculations have real cost consequences because they must be carried out by some entity actually residing in the world. ACE[27] modelling forces the modeller to respect this constraint. An ACE model is essentially a collection of algorithms (procedures) that have been encapsulated into the methods of software entities called 'agents'. Algorithms encapsulated into the methods of a particular agent can only be implemented using the particular information, reasoning tools, time and physical resources available to that agent. This encapsulation into agents is done in an attempt to achieve a more transparent and realistic representation of real world systems involving multiple distributed entities with limited information and computational capabilities."

2.6.3 What It Is and Is not

So agent-based modelling is a method, or approach, which examines the interactions of "things" or "entities" rather than a particular thing or collection of things to be replicated. While modellers may have an instinct for what is or is not an agent-based model, the reality is that it falls, along with several related fields that also focus on interacting things, on a spectrum with fuzzy boundaries and confusing overlaps. Before proceeding to examine our agents, it is important to differentiate where it lies on the spectrum in relation to other "thing-centric" fields, and whether the distinctions between the concepts are useful for a given investigation.

Agent-Based Model *What happens when ...?* Agent-based models are constructed to discover possible emergent properties from a bottom-up perspective. They attempt to replicate, in silico, certain concepts, actions, relations or mechanisms that are proposed to exist in the real-world, in order to see what happens. Generally, agent-based modelling has no desired state or task to be achieved, instead merely describing the entities and observing how they interact in order to

---

[27] Agent-Based Computational Economics; essentially agent-based modelling with agents containing economic decision models.

explore the system's possible states. An agent-based model can examine how farmers might adapt to climate change (Schneider et al. 2000), the co-evolution of autocatalytic economic production and economic firms (Padgett et al. 2003) or the behaviour of an abstract economy (Kauffman 2008). The model acknowledges that reality consists of many components acting, relatively autonomously, in parallel, and that no specific predictions can be made, but that patterns, tendencies, and frequent behaviours shown in the model may be relevant to the real world. While an agent-based model generally has no set state to achieve, replicating some real world phenomenon to a desired degree of accuracy means that some models become less about seeing what happens and more about seeing what it takes to make something specific happen.

Multi-agent System *How can I make a … ?* Multi-agent systems are often, but incorrectly, used interchangeably with agent-based modelling because they also use discrete, parallel, autonomous components (or agents) to examine system emergence. The main difference is that agent-based modelling sets up agents believed to have crucial characteristics of real world analogs to see what happens when they do whatever they do, while in a multi-agent system agents are set up with exactly the characteristics, connections and choices that they need to achieve certain desired emergent states. For example, it be used to develop process design using component collaboration and local information (Hadeli et al. 2004), the design of an advanced e-commerce agent (Lee 2003), a predictive control system for transportation networks (Negenborn et al. 2006), and the design of cooperative agents in a medical multi-agent system (Lanzola et al. 1999). A multi-agent system is a an attempt to control emergent problems, like traffic control or agenda synchronisation, that are not best solved by top down approaches but must resolve all conflicts (i.e. no traffic jams or conflicting appointments). While usually trying to solve a given problem rather than replicate the behaviour of actors in real world situations, it can sometimes look quite a lot like an agent-based model if they problem to be solved involves exploring the unpredictable behaviour of human-like agents.

Artificial Intelligence *Can he do this … ?* Artificial intelligence can be seen as zooming in on the agent. Consciousness, learning, object detection and recognition, decision making and many other facets of intelligence can be considered emergent properties and artificial intelligence researchers are attempting to replicate these, much as agent-based models seek to replicate the emergent properties of industries, economies and cultures. Although often studied in isolation, groups of artificial intelligence agents, usually called distributed artificial intelligence, would be a return to the level of zoom that allows for emergent properties between agents instead of only within agents. This would be almost indistinguishable from a multi-agent system if the distributed artificial intelligence agents were trying to solve a particular problem or achieve a certain state, as when teams of intelligent, problem solving robots try to play a game of football. Alternatively, if distributed artificial intelligence agents are instead left to their own devices while researchers observe their output, like any communicative systems they might develop, then they start to look a lot like an agent-based model (Honkela and Winter 2003).

Object-Oriented Program? Rohilla Shalizi (2006) states that:

> "While object-oriented programming techniques can be used to design and build software agent systems, the technologies are fundamentally different. Software objects are encapsulated (and usually named) pieces of software code. Software agents are software objects with, additionally, some degree of control over their own state and their own execution. Thus, software objects are fixed, always execute when invoked, always execute as predicted, and have static relationships with one another. Software agents are dynamic, are requested (not invoked), may not necessarily execute when requested, may not execute as predicted, and may not have fixed relationships with one another."

In essence, agents may be built with object oriented programming software, and when given very simple rules and limited behavioural options, they behave very like objects. But at heart, agents are designed to be unlike normal objects because they flagrantly ignore the usual programming goal of eliminating repetitive or unnecessary elements. By having multiple and similar agents or components, many of which may not have any actions or whose actions seem ineffective, pointless, counterproductive or irrelevant, a simulation using agents cannot be elegant, streamlined or minimal code. However, as with all bottom-up approaches, the messy, repetitive, unexpected relations are important, and the surprisingly concise results and solutions can only be seen as more than the sum of the parts.

2.7 Anatomy of an Agent-Based Model

Our dissection of an agent-based model begins with a schematic overview, presented in Fig. 2.6, followed by a detailed description of the Agent, its states and its behaviour rules, before a look at the Environment. Finally, we detail the structure and organisation of agent interactions and aspects of time. While very theoretical at this stage, these notions will become concrete in Chap. 3, which discusses in detail the process of creating a model with the anatomy described here.

2.7.1 Agent

Agents are reactive, proactive, autonomous and social software entities, a computer program or " an encapsulated computer system that is situated in some environment, and that is capable of flexible, autonomous action in that environment in order to meet its design objectives" (Jennings 2000). Agents are:

1. Encapsulated, meaning that they are clearly identifiable, with well-defined boundaries and interfaces;
2. Situated in a particular environment, meaning that they receive input through sensors and act through effectors;
3. Capable of flexible action, meaning that respond to changes and act in anticipation;
4. Autonomous, meaning that they have control both over their internal state and over their own behaviour; and

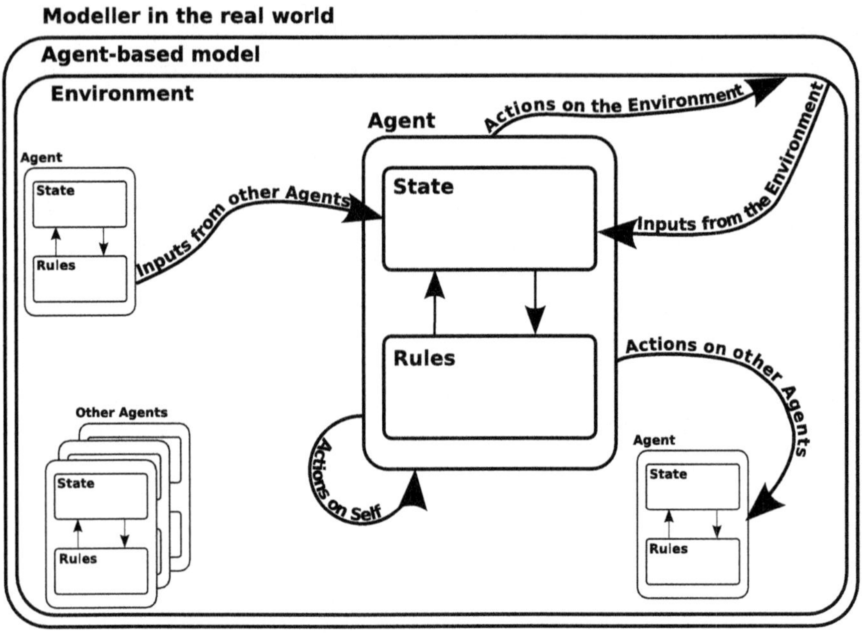

Fig. 2.6 Structure of an agent-based model

5. Designed to meet objectives, meaning that they attempt to fulfil a purpose, solve a problem, or achieve goals.

(adapted from Jennings 2000).

The agent is the smallest element of an agent-based model, the atomic element of a generative theory, and some would even say that the "agent is the theory". An agent is able to perform actions on itself and other agents, receive inputs from the environment and other agents, and behave flexibly and autonomously because, as shown in Fig. 2.6, an agent consists of both states and rules.

2.7.1.1 State

An agent's state is the specific collection of parameters that defines an agent (Wooldridge and Jennings 1995), or all of the relevant information about what this agent is at this moment. The internal, local and global states of each agent, any of which can be static or dynamic, contribute to its overall state.

The internal state belongs to the agent and only the agent, and covers the current values of all the possible properties that an agent could possibly have. An agent representing a greenhouse would have an internal state composed of all the values for current growing capacity, energy use, owner, and financial balances, among many other possible properties, while a light switch agent would have one of two possible internal states: on and off. The internal state can be private, public, or a mixture, if only some properties are observable by other agents or if a property is observable by only some other agents.

However, an agent's actions are also dependent on the actions and inputs from others with which it interacts. Thus, the local state consists of the internal state (private and public) plus all of the publicly observable states of the agents that our agent is interacting with. This puts the internal state into a context and gives some sense to the values, allowing the agent to act based on not only its internal state but the relationship that internal state has to the immediate surroundings.

Finally, the global state is comprised of the internal state, the local state and all of the relevant states in the whole of the observable or influencing environment. With these three states, you can already see how every agent is a complex system, embedded in nested networks of influence that act on various time scales and levels of interaction. The agent uses its internal, local and global states as the basis for applying behavioural rules in order to produce actions.

2.7.1.2 Changing States

Rules Rules, or the "internal models" (Holland 1996) of agents describe how states are translated to actions or new states. Rules should be understood as mechanical decision rules or transformation functions, rather than the more colloquially used social notions of rules as regulations or agreements.

The rules of agents in models of complex systems are usually based on an assumption of rationality or bounded rationality (Simon 1982). For example, a common decision rule might be that agents attempt to maximise some utility, but the agent may or may not have access to information about the other agents with which they interact, may or may not be able to record the outcome of previous actions in order to learn, or may have limits on the computation allowed to process any information in order to mimic the limits that human decision makers face. Decision rules specify what an agent will do with the information that they have access to, as well as how they will perform any actions. Rules can be static or dynamic, and may depend on the internal, local and environmental states. Importantly, agents could choose not to perform an action, either because the rules allow for inaction or because the rules call for probabilities, noise or random elements that alter the normal actions. There are several types of decision rules often used in agent-based modelling. These are:

Rule based Rule based decision rules are the most common, usually in the form of nested if-then-else structures. These are the most common type of decision rules, as they are very easy to implement, and directly couple observed behaviour into decision structures.

Multi-criteria decision making Another common technique used for agent decision making is multi-criteria decision making. It is a technique that allows different choice options to be compared, for example by assigning weights, enabling the agents to have preferences or probabilities. For example, a greenhouse agent might weight the emissions of a CHP unit more heavily than the price when making purchasing decisions, among other factors, resulting in purchase choices that are not readily obvious.

Inference engines Also known as expert systems, inference engines take facts (states) and decision heuristic to construct a decision tree in order to reach conclusions about which action should be taken. Such systems are often used when an agent needs to base a decision on a lot of real world data, and are often found in engineering, financial and medical applications.

Evolutionary computing When agents need to find a optimal solution in a very complex or large solution space, genetic algorithms can be employed. Agents generate a large number of solutions, evaluate their fitness against a fitness function, select a group of "best" solutions and apply genetic recombination on them in order to make better ones. Such techniques can be very computationally expensive.

Machine learning Neural networks can be used when an agent needs to make decisions based on patterns. Neural networks work as classifier systems, allowing the agent to determine into which category a observed pattern falls, and therefore decide which action is appropriate. Neural networks are also computationally intensive and may require a training period, to allow the agent to learn the categories and which actions are best for each category, before any decisions can be used.

Actions Actions are the actual activities that agents perform based on the application of decision rules on their states. For example, if a light switch agent has a rule that says "If state = off and time = 8:00, set state = on". Thus, the rule uses the agent's own internal state (off or on), a global state (the time) and the decision rule to perform an action that changes its own internal state.

Actions can also be directed at other agents. For example, a tax collecting agent might have a rule such as "For each greenhouse grower agent, if yearly profits exceed 10000, deduct 10 %". This agent then would consider none of his own internal states, but would look at the states of other agents (local), some sort of calendar (global) and act directly on the financial balances of the grower agents (the internal state of another agent).

Of course, agents can also *not* act, which can be understood as an action as well.

Behaviour The agent behaviour is the overall observable sum of the agent's actions and state changes. It is an emergent property caused by the interaction of the internal, local and environmental states and the decision rules. Overall system (or model) behaviour is an emergent property of the interactions between all of the agents behaviours and the environment.

Greenhouse Example This incomplete example of the agent anatomy of a greenhouse model is drawn from a complete model of greenhouses developed in a stepwise fashion in Chap. 3.

In this model, there is only one type of agent, representing a greenhouse grower. These agents have internal states composed of the values of various properties, such as what technologies they currently own, what opinions they hold about those technologies, how much money they have, how much profit they earned last season, and what kind of crops they produce. Some of these, such as what technologies they own, are public, while others, such as the exact amount of money owned, are private. The local state also contains the neighbours with whom the agent has communicative links, and the publicly available properties of those agents, such as what technologies they own. The global environment further consists of some environmental properties, such as the price of electricity.

The greenhouse agents have rules that govern how they purchase technologies and how they form opinions about them. For example, agents have a rule like "When a currently owned technology expires, purchase a new technology in the same technology class which has the highest opinion and which is lower in cost than the current money owned." Thus, when a greenhouse agent's heater breaks, they purchase a replacement heater, and that heater is the best they can afford, according to their own opinions of which technologies are best.

The actions an agent can take include selling crops, buying technologies, updating account balances, and updating their opinions of technologies. The agent behaviour includes not only the actions they take, but also the changes in technologies, opinions and account balances over the course of the simulation run.

2.7.2 *Environment*

Agents must be somewhere, and that somewhere gives the agents input through sensors and receives the output or effects of the agent action. This somewhere, in which the agents "live", is the environment, and it contains all the information external to the agent used in the decision making processes and provides a structure or space for agent interaction. The environment contains everything, including other agents, that affects an agents, but which is not the agent itself, which means of course, that an agent's environment is context dependent.[28] Agents can affect the environment and be affected by it as a consequence of the specific rules they use for actions.

2.7.2.1 Information

Because the environment, in a strict sense, is context dependent, we will use a looser, less context dependent definition here. Rather than consider each agent's environment uniquely to see how they differ, much the way internal states differ between agents, we focus on the similarities, so that the environment is a shorthand for the information, structure and goings on in the global states. Thus, we can say that the

[28] Ah, our old friend, context dependency. In this case, it means that no two agents will have exactly the same environment, because every agent will be in the environment for other agents, but not for himself.

environment provides all the things an agent needs to know, and all the ways to do the things that it does, and that are not contained in the agent itself or in its immediate neighbours. The environment will have some elements provided by the model itself, while others could be set by the modeller, or can be emergent.

Environmental information provided by the model itself tends to be quite dull, although crucial. Things like the passage of time, which is so important that we discuss in Sect. 2.7.3 below, is an environmental aspect provided by the model that agents use to make decisions. Many rules use elements of time to make decisions or dictate actions. For example, an agent might have a rule that says "Every turn, make and sell a variety of products. After 10 turns, review the best and worst selling products, adjust the probability of making those products, eliminate the worst selling and introduce a brand new product." The environment provides the measure of how many turns, or time ticks, have passed, providing information on whether this turn is a turn for making and selling, or for the added action of reviewing sales and production.

When an aspect of the environment is provided by the modeller, it can be considered a model parameter. These parameters can be static or dynamic, and whole sets of these parameters are often considered scenarios that the modeller is interested in. For example, agents might use some global variable, such as temperature, to make decisions. The modeller can set value of the temperature to make a static, global variable. If the modeller runs the model several times, each time with a different temperature, then each model run represents a scenario, or experimental condition under testing. The modeller could also design the model so that the temperature is dynamic, either varying across a set range, following historical data that has been fed into the model, or determined by a probability function. Different ranges, different sets of data, or different probability functions would represent different scenarios for experimentation.

Aspects of the environment could be provided by emergent properties of the model itself rather than directly by the modeller. Emergent properties arise through interaction between agents in the simulation, and may or may not be directly accessible to individual agents, depending on how they must be calculated and whether or not the calculations rely on private or public states. For example, market prices are emergent, and no single agent can determine the prices individually, because agents cannot usually access the supply or demand of particular products, nor the willingness of other agents to pay a given price for a product. Nevertheless, the distributed effects of many private actions and calculations results in a specific price per product, and agents may use that price to determine actions.

2.7.2.2 Structure

The environment also provides the structure in which the agents are situated, which can also be static or dynamic and which can be set by the modeller or derived as an emergent property of the model. If it is not relevant to the decision making rules, agents need not know anything about that structure because the environment will provide only the information needed by the agents when they need it. For example, a rule might be "Ask all your neighbours if they have any tomatoes to sell. If they do, buy the cheapest". The agent can simply ask the environment "Who are my neighbours?", and then use the answer to ask those other agents about their tomato availability. If it does not matter to his decision rules, then he will never know if he has more or fewer neighbours than another agent or if the neighbours are the same as the last time he asked for tomatoes.

Although the agent may not know what structure in which he is situated, that structure can significantly affect the model performance. In agent-based modelling, we can distinguish four main types of structures investigated so far, a unstructured soup, a regular space, a small-world network and a scale-free network. The different structures and network topologies each share some characteristics with human social networks, which may or may not be important, but also influence many things, from the time it takes the simulation to reach some important behaviour to the amount of processor power that is needed to run a simulation for a given number of steps. Furthermore, multidimensional networks are possible so that a set of agents can be connected in more than one way at the same time so that communications flow along some connections, money along others, physical materials along yet others, etc. Choosing one structure over another, or multiple structures, allows the modeller to bring more complexity and realism to the model, an important part of modelling a complex adaptive systems. Nevertheless, the more complex and realistic the structure, the more difficult it is to statistically analyze and the more likely it is to obscure any relations in the model.

Soup A popular and statistically easy to analyze structure is the mean-field organisation, or "soup", which covers completely random organisations (i.e. agents are equally likely to interact with all other agents) or fully connected organisations (i.e. agents interact with all other agents). A limited number of agents, or a high number of interactions, means that the agents will be fully connected, while a large number of agents, fewer interactions, or a high degree of agent turnover would lead to random interactions, potentially affecting model behaviour. Soups have a short average path length, meaning that there are few steps needed to connect any two agents, which is also true of real-life human networks. But they have a low clustering coefficient, meaning that the there are no subsets of agents that interact more frequently with each other than with other agents, which is not true of human social networks. Soups tend to proceed quite rapidly, reaching plateaus, converging on behaviours, or running until stop conditions are reached much faster than other structures, although they typically use more memory resources as well.

Space Another classic structure for organising the agents is the space or regular structure, in which agents are connected to a set of neighbours but not to all the other agents. They are usually arranged in a regular pattern such as a square or hexagonal grid which may be cast onto a toroid to prevent edge effects. Agents can also be situated in a more physically defined space within a GIS map, where neighbourhood is defined in terms of actual distance, a common approach in fields such as spatial planning and geography. This structure provides a sense of close and far between agents, and because neighbouring agents are connected to the most of the same agents as their neighbours, this structure displays a high clustering coefficient, as do human networks. The high clustering means that local subsets of agents reach convergence or display coherent behaviours quickly. This rapid local convergence can lead to slow global convergence, but reduces the computing power or memory needed to run the simulation. Further, as spaces lend themselves to analysis with the usual methods of statistical mechanics (Baronchelli et al. 2005, 2006) they have been a popular starting point for many researchers.

Small-World Networks Discovered more recently, small-world networks start with a regularly structured space and randomly replace a small number of the local connections with long distance connections. After a very small number of rewirings, the network takes on the short average path length characteristic, and fast global convergence times, of soups without sacrificing the high clustering coefficient, and efficient resource use, of spaces. Despite being discovered quite recently, small-world networks have been subject to much investigation, especially as they relate to regular and random networks. By including both a short average path length and a high clustering coefficient, small-world networks are more realistic than either soups or spaces, but are more difficult to analyze statistically.

Scale-Free Networks Scale-free networks have high clustering coefficients and short average connection length, as do small-world networks, but they also have "hubs" that can dramatically speed up the timescales in the model. These hubs stem from a power law degree distribution,[29] which distributes connections, probability to interact, or popularity among the agents such that a very few agents are highly connected while the vast majority have few connections, matching real-life human social networks. The scale-free networks show sharp transitions toward convergence, as do real life populations, and they reach convergence almost as quickly as soups, while using memory resources efficiently. Other structures can be converted to a scale-free network by incorporating growth with some sort of preferential attachment strategy, so scale-free networks can be used in dynamic structure models. While scale-free networks have importantly realistic traits and behaviours, the presence of hubs can introduce problems with robustness and instability if not well considered. For example, if only a few agents are chosen at random to be active each turn, then the overwhelmingly numerous non-hub agents will tend to be selected. If those agents then select one of their neighbours as the recipient of the action (reactive agent), then the hub-agents will tend to be chosen as the recipient quite often, because they are in the set of neighbours of almost every other agent. Thus, depending on the action taken, the hub agents will have experiences and behaviours different than other agents, even if they share all the same decision rules, and may display quite different behaviour. It is not unrealistic, but the implications are not

[29]Random interaction soups and small-world networks only have a Poisson degree distribution, meaning that extremely popular agents are as common as extremely unpopular agents.

trivial as the divergent behaviour may speed up convergence, promote robustness, or spread instability, depending on how the model works.

Greenhouse Example The environment in the greenhouse model includes some information provided by the model, most notably, time. This is complicated and important enough to have it's own greenhouse example later, so we now move on directly to detail the other informational and environmental factors in this example. Although not directly used as input to the decision making rules in this model, the environment provides the agents with a social structure that they use to interact. The greenhouse growers are placed in a scale free network that mimics the structure of natural human social networks. A few agents are highly connected and acts as hubs for information about opinion on technology, while the majority of agents are in the periphery of the network with a limited number of neighbours. Were it allowed to vary, the structure of the agent interactions would be a model parameter. For this model, however, other structures were not deemed to be realistic scenarios.

Some of the model parameters that can vary, and which are also sources of information from the environment, include how many agents are in the simulation, how big the greenhouse sizes can be, and with what probability the agents are likely to heed the advice of their neighbours when making technology purchase decisions. One possible scenario might be the case of few greenhouse grower agents, ranging in size from 1 to 25 hectares, and with all neighbours opinions valued highly. Another scenario might be many growers, with greenhouse size ranging from 10 to 20 hectares, and with growers paying very little attention to the opinions of their neighbours.

The environmental information also includes information that is not determined by the model itself, nor given as a parameter by the modeller, but which is an emergent property of the model behaviour. In this greenhouse example, agents compare their own profit to the average profit made by all greenhouse in order to determine how satisfied they are with their technology choices. As growers do not have access to the profit of their neighbours, much less to the profit of all the greenhouse in the simulation, they rely on the environment to calculate the average profit from the private information of all growers.

2.7.3 Time

The final aspect of agent-based model that needs to be discussed is the issue of time. Time can be considered a part of the environment, but is so ubiquitous that it demands special attention and unique considerations. Real world complex adaptive systems take place in a continuous real time, and with elements truly acting in parallel. If we are to satisfy Ashby's requirement, we must ensure that these aspects are properly represented in a model, and we must understand them well if we are to represent them.

Discrete Time While reality takes place in real, continuous, time, agent-based models are forced to happen in the discrete time of computers. All conventional computers work with timed instruction clocks, performing rounds of operations within each time step. This reflected by the use of a *tick* as the smallest unit of time. Simulations can play with discrete time by redefining how much time a tick is meant to represent, with no theoretical lower or upper limits. However, if the time needed to compute a single tick is longer than the amount of real time that tick is meant to represent, then this places a particular practical limit on one way that simulations are often used. A simulation that runs slower than reality is not very useful for predictions.

Assumption of Parallelism While the discrete time is significantly different from reality, the main problem is parallelism. Although real world complex adaptive systems are massively parallel, only very recently have multi-core computers enabled the performance of more than one task at a time.[30] In order to represent the parallelism of the real world with a serial processing device, all actions are scheduled to occur one after the other, but are assumed to happen at the same time. The disjoint between what actually happens and what is assumed to happen can create significant problems. For example, when simulating bird flocks, each bird constantly observes its neighbours speed and position and adjusts to them. However, one bird has to go first, and he can only observe the *previous* states of all other birds in order to decide how to move itself. Then next bird to move observes the *current* state of the first bird, and the *previous* state of the other birds, and so on through all the birds. One option to deal with this might be that the birds always move in the same order, so that each bird at least has an internally consistent relation between his observations and actions, although his observation-action relation will be unlike that of any other bird. Alternatively, the birds could observe and act in random order, so that no two turns in a row will have the same observation-action relationship, but these will be roughly consistent between all birds. Or the modeller can choose that a tick takes place in two parts, first observing the position of all other birds, and then moving to where they think they need to be. These decisions are not trivial, as they can provide the agents a "glimpse into the future" or purposefully restrict information by only allowing access to the inherently outdated. The explicit management of the order of agent interaction *over* and *within* time is performed by the scheduler.

Scheduler Schedulers are the central controller of a system that *simulates* a system without any central control.[31] The scheduler progresses the ticks and ensures that all "parallel" actions are executed. Most commonly, this involves randomising the iteration order of agents at each step. If not properly randomised or otherwise controlled, first-mover advantage modelling artifacts may appear. For example, a greenhouse agent has to observe a market, find the best price, and sell all of his

[30] Multitasking on single core computers is achieved by very quickly switching between tasks.

[31] Finally! Now we know the dirty little secret of agent-based modelling ....

produce. If the same agent always goes first, then he will always be able to sell his entire stock at the best price that could possibly be offered, giving him a distinct advantage over the poor agent who always goes last and has to settle for whatever nobody else wanted. Depending on the exact implementation of the modelling software, the scheduler might allow for very fine grained control of specifying the order of particular actions of particular agents, making sure that for example market clearing agent always goes last, or that the order of actions of a particular agent are randomised.

Greenhouse Example In the greenhouse model, each tick represents one year. At each tick, agents perform some actions, such as selling produce and balancing their books, and the environment "ages" the technologies. The agents perform other actions, such as buying replacement technologies, only when triggered by one of their technologies reaching its expiration age. Thus, the agents use information directly from the model in the form of ticks, information indirectly from the model, in the form of their technology ages, and their own decision rules to act in time.

Furthermore, each time tick is composed of several phases that all agents perform in a random order before moving on to the next phase. The first phase sees all agents sell their produce and calculate their own profit, at which point the environment calculates the average profit for the round. In the next phase, agents compare their own profit to the average and use the difference to form opinions on the technologies that they own (more profit than the average means they form a positive opinion of their technologies, lower than average means they form poor opinions). After that, agents share their current opinions with neighbours, before incorporating the opinions of their neighbours back into their own opinion of the technologies. Finally, agents check to see if any technologies have expired, and if so they use their current account balance and their newly updated opinions to purchase a replacement.

Importantly, the agents all form their own opinion before sharing any opinions with neighbours. If some agents shared their opinions based solely on their profit compared to the average before other agents got the chance to compare the profits and form an opinion, then the opinions of the first agents to speak would be more influential than later agents. The opinions of the first agents would influence others, who would then repeat their compromised opinions to others, influencing them. If this were totally random, perhaps it would all balance out to prevent first-mover advantage, but by separating the actions of opinion formation, sharing and reconsideration, no opinion has any undue influence from first-mover advantages.

In this model, the representation of time as multiple phases of action per "one year" tick cannot change. Other simulations might give the option to vary the representation of time, the order of interactions, or the representation of parallelism, and thus would be model parameters as part of a scenario. The possibility of varying such fundamental model details as the conceptualisation of time might be considered unique formalisations. However, to truly be a distinct formalisation, any possible variance would need strong justification and thorough theoretical support, so slapping a variable on for time is no substitute for proper multi-formalism.

References

Aldrich, H., & Whetten, D. (1981). Organization-sets, action-sets, and networks: making the most of simplicity. In *Handbook of organizational design* (Vol. 1, pp. 385–408).

Alexander, C. (1973). A city is not a tree. *Surviving the city: a sourcebook of papers on urban livability*, p. 106.

Allen, T., Tainter, J., & Hoekstra, T. (1999). Supply-side sustainability. *Systems Research and Behavioral Science, 16*, 403–427.

Argyris, C., & Schon, D. A. (1996). *Organisational learning II; theory, method and practice*. Amsterdam: Addison-Wesley.

Ashby, W. R. (1968). Variety, constraint, and the law of requisite variety. In *Modern systems research for the behavioral scientist*. Chicago: Aldine.

Axelrod, R. (1980). More effective choice in the prisoner's dilemma. *Journal of Conflict Resolution, 24*(3), 379–403.

Baronchelli, A., Dall'Asta, L., Barratt, A., & Loreto, V. (2005). Topology induced coarsening in language games. *Physical Review E 73*, 015102.

Baronchelli, A., Loreto, V., Dall'Astra, L., & Barratt, A. (2006). Bootstrapping communication in language games: strategy, topology and all that. In *Proceedings of the 6th international conference on the evolution of language* (pp. 11–18).

Bijker, W., Hughes, T., & Pinch, T. (1987). *The social construction of technological systems: new directions in the sociology and history of technology*. Cambridge: MIT Press.

Boer, C., Verbraeck, A., & Veeke, H. (2002). Distributed simulation of complex systems: application in container handling. In *Proceedings of SISO European simulation interoperability workshop* (pp. 24–27).

Bonen, Z. (1981). Evolutionary behavior of complex sociotechnical systems. *Research Policy, 10*(1), 26–44.

Borshchev, A., & Filippov, A. (2004). From system dynamics and discrete event to practical agent based modeling: reasons, techniques, tools. In *Proceedings of the 22nd international conference of the system dynamics society* (pp. 25–29).

Box, G. (1979). Some problems of statistics and everyday life. *Journal of the American Statistical Association, 74*(365), 1–4.

Boyson, S., Corsi, T., & Verbraeck, A. (2003). The e-supply chain portal: a core business model. *Transportation Research, Part E, 39*(2), 175–192.

Buchanan, M. (2000). In *Ubiquity: the science of history or why the world is simpler than we think*. London: Weidenfeld.

Buchanan, M. (2009). Economics: Meltdown modelling. *Nature, 460*(7256), 680.

Burdulis, A., Fitz, W., Vargas-Voracek, R., Lang, P., Steines, D., & Tsougarakis, K. (2010). *Surgical tools facilitating increased accuracy, speed and simplicity in performing joint arthroplasty*. US Patent App. 12/776,701.

Burks, A. (1970). *Essays on cellular automata*. Champaign: University of Illinois Press.

Callaway, D., Newman, M., Strogatz, S., & Watts, D. (2000). Network robustness and fragility: percolation on random graphs. *Physical Review Letters, 85*, 5468.

Campbell, N. (2002). *Biology*. Redwood City: Benjamin Cummings.

Checkland, P., & Checkland, P. (1999). *Systems thinking, systems practice: includes a 30-year retrospective*. New York: Wiley.

Cohen, M., March, J., & Olsen, J. (1972). Garbage can model of organizational choice. *Administrative Science Quarterly, 17*(1), 1–25.

Colletta, L. (2009). Political satire and postmodern irony in the age of Stephen Colbert and Jon Stewart. *The Journal of Popular Culture, 42*(5), 856–874.

Conway, J. (1970). The game of life. *Scientific American, 223*(4), 4.

Corsi, T., Boyson, S., Verbraeck, A., Van Houten, S., Han, C., & Macdonald, J. (2006). The real-time global supply chain game: new educational tool for developing supply chain management professionals. *Transportation Journal, 45*(3), 61.

Coulson, J., & Richardson, J. (1999). *Coulson & Richardson's chemical engineering*, Stoneham: Butterworth/Heinemann.

Crutchfield, J. (1994). The calculi of emergence: computation, dynamics and induction. *Physica D*, *75*(1–3), 11–54.

Darwin, C. (1985). *The origin of the species*. Baltimore: Penguin.

David, P. (2000). Path dependence and varieties of learning in the evolution of technological practice. In *Technological innovation as an evolutionary process* (p. 119). London: Cambridge University Press.

Dawkins, R. (1990). *The selfish gene*. London: Oxford University Press.

DeLaurentis, D., & Crossley, W. (2005). A taxonomy-based perspective for systems of systems design methods. In *2005 IEEE international conference on systems, man and cybernetics* (Vol. *1*).

Dennet, D. (1996). *Darwin's dangerous idea: evolution and the meanings of life*. New York: Simon & Schuster.

Economides, N. (1996). The economics of networks. *International Journal of Industrial Organization*, *14*(6), 673–699.

Economist, T. (2010). Agents of change. *The Economist*. http://www.economist.com/node/16636121.

Epstein, J. (1999). Agent-based computational models and generative social science. *Complexity*, *4*(5), 41–60.

Farmer, J., & Foley, D. (2009). The economy needs agent-based modelling. *Nature*, *460*(7256), 685–686.

Ferguson, C. (1968). Absence of copula and the notion of simplicity: a study of normal speech, baby talk, foreigner talk and pidgins.

Foerster, H. (1972). Perception of the future and the future of perception. *Instructional Science*, *1*(1), 31–43.

Forrester, J. W. (1958). Industrial dynamics—a major breakthrough for decision makers. *Harvard Business Review*, *36*(4), 37–66.

Forrester, J., & Wright, J. (1961). *Industrial dynamics*. Cambridge: MIT Press.

Funtowicz, S., & Ravetz, J. (1993). Science for the post-normal age. *Futures*, *25*(7), 739–755.

Futuyma, D. (1983). Evolutionary interactions among herbivorous insects and plants. In *Coevolution* (pp. 207–231). Sunderland: Sinauer.

Gaukroger, S. (2001). *Francis Bacon and the transformation of early-modern philosophy*. Cambridge: Cambridge University Press.

Geels, F. W. (2004). From sectoral systems of innovation to socio-technical systems—Insights about dynamics and change from sociology and institutional theory. *Research Policy*, *33*(6–7), 897–920.

Gleick, J. (1997). *Chaos: making a new science*. New York: Vintage/Ebury.

Gordon, G. (1978). The development of the general purpose simulation system (GPSS). In *History of programming languages I table of contents* (pp. 403–426).

Green, P. (1981). A new look at statistics in fission-track dating. *Nuclear Tracks*, *5*(1), 77–86.

Hadeli, W., Valckenaers, P., Kollingbaum, M., & Brussel, H. V. (2004). Multi-agent coordination and control using stigmergy. *Computers in Industry*, *53*(1), 75–96.

Hartmanis, J., Sewelson, V., & Immerman, N. (1983). Sparse sets in np-p: exptime versus nexptime. In *STOC '83: proceedings of the fifteenth annual ACM symposium on theory of computing* (pp. 382–391). New York: ACM.

Hietbrink, O., Ruijs, M., & Breukers, A. (2008). *The power of dutch greenhouse vegetable horticulture: an analysis of the private sector and its institutional framework*. Technical report 2008-049, LEI Wageningen UR, The Hague.

Hix, J. (1996). *The glasshouse*. London: Phaidon Press.

Holland, J. (1996). *Hidden order; how adaptation builds complexity*. Reading: Addison-Wesley.

Holling, C. S. (2001). Understanding the complexity of economic, ecological, and social systems. *Ecosystems*, *4*(5), 390–405.

Honkela, T., & Winter, J. (2003). Simulating language learning in community of agents using self-organizing maps.

Hume, D. (1962). *A treatise of human nature*, vol. *1*. Glasgow: Collins.
Iyengar, S., & Kamenica, E. (2010). Choice proliferation, simplicity seeking, and asset allocation. *Journal of Public Economics*, *94*(7–8), 530–539.
Jablonka, E., & Ziman, J. (2000). Biological evolution: processes and phenomena. In *Technological innovation as an evolutionary process* (pp. 13–26). Cambridge: Cambridge University Press.
Janis, I. (1982). *Groupthink: psychological studies of policy decisions and fiascoes*. Boston: Houghton.
Jantzen, D. (1980). When is it coevolution. *Evolution*, *34*, 611–612.
Jennings, N. (2000). On agent-based software engineering. *Artificial Intelligence*, *117*(2), 277–296.
Jones, R. (1965). The structure of simple general equilibrium models. *The Journal of Political Economy*, *73*(6), 557.
Katz, J. S. (2006). Indicators for complex innovation systems. *Research Policy*, *35*(7), 893–909.
Kauffman, S. (2008). *Reinventing the sacred: a new view of science, reason and religion*.
Kauffman, S., & Johnsen, S. (1991). Coevolution to the edge of chaos—coupled fitness landscapes, poised states, and coevolutionary avalanches. *Journal of Theoretical Biology*, *149*(4), 467–505.
Kay, J. (2002). On complexity theory, exergy and industrial ecology: some implications for construction ecology. In C. Kibert, J. Sendzimir, & B. Guy (Eds.), *Construction ecology: nature as the basis for green buildings* (pp. 72–107). London: Spon.
Kellert, S. (1993). In *the wake of chaos: unpredictable order in dynamical systems*. Chicago: University of Chicago Press.
Kim, J. (1999). Making sense of emergence. *Philosophical Studies*, *95*(1), 3–36.
Lanzola, G., Gatti, L., Falasconi, S., & Stefanelli, M. (1999). A framework for building cooperative software agents in medical applications. *Artificial Intelligence in Medicine*, *16*(3), 223–249.
Lee, R. (2003). Ijade surveillant—an intelligent multi-resolution composite neuro-oscillatory agent-based surveillance system. *Pattern Recognition*, *36*(6), 1425–1444.
Leontief, W. (1998). Environmental repercussions and the economic structure: an input-output approach. *International Library of Critical Writings in Economics*, *92*, 24–33.
Lindblom, C., Cohen, D., & Warfield, J. (1980). Usable knowledge, social science and social problem solving. *IEEE Transactions on Systems, Man and Cybernetics*, *10*(5), 281.
Luhmann, N. (1995). *Social systems*. Stanford: Stanford University Press.
Mandelbrot, B. (1983). *The fractal geometry of nature*. New York: Freeman.
Mandeville, B., & Harth, P. (1989). *The fable of the bees*. Baltimore: Penguin.
Mikulecky, D. (2001). The emergence of complexity: science coming of age or science growing old? *Computers and Chemistry*, *25*(4), 341–348.
Miller, D. (1993). The architecture of simplicity. *The Academy of Management Review*, *18*, 116–138.
Morin, E. (1999). Organization and complexity. *Tempos in Science Nature: Structures, Relations, Complexity*, *879*, 115–121.
Munger, M. (2008). Blogging and political information: truth or truthiness? *Public Choice*, *134*, 125–138.
Negenborn, R., De Schutter, B., & Hellendoorn, H. (2006). Multi-agent model predictive control of transportation networks. In *Proceedings of the 2006 IEEE international conference on networking, sensing and control (ICNSC 2006)* (pp. 296–301).
Newman, M. (2003). The structure and function of complex networks. *SIAM Review*, *45*, 167–256.
Ottens, M., Franssen, M., Kroes, P., & van de Poel, I. (2006). Modelling infrastructures as sociotechnical systems. *International Journal of Critical Infrastructures*, *2*(2–3), 133–145.
Padgett, J., Lee, D., & Collier, N. (2003). Economic production as chemistry. *Industrial and Corporate Change*, *12*(4), 843–877.
Prigogine, I. (1967). *Introduction to thermodynamics of irreversible processes* (3rd ed.). New York: Interscience.
Prigogine, I., & Stengers, I. (1984). *Order out of chaos: man's new dialogue with nature*. Boulder: New Science Library.

Reynolds, C. (1987). Flocks, herds and schools: a distributed behavioral model. *Computer Graphics*, *21*, 25–34.
Roberts, R. (1989). Serendipity: accidental discoveries in science. In *Serendipity: accidental discoveries in science* (p. 288).
Rohilla Shalizi, C. (2006). Methods and techniques of complex systems science: an overview. In *Complex systems science in biomedicine* (pp. 33–114). Berlin: Springer.
Rosenberg, R., & Karnopp, D. (1983). *Introduction to physical system dynamics*. New York: McGraw-Hill.
Ryan, A. (2008). What is a systems approach? arXiv:0809.1698.
Schelling, T. (1971). Dynamic models of segregation. *The Journal of Mathematical Sociology*, *1*(2), 143–186.
Schneider, S., Easterling, W., & Mearns, L. (2000). Adaptation: Sensitivity to natural variability, agent assumptions and dynamic climate changes. *Climate Change*, *45*(1), 203–221.
Schonberger, R. (1982). *Japanese manufacturing techniques: nine hidden lessons in simplicity*. New York: Free Press.
Simon, H. (1982). *Models of bounded rationality*. Cambridge: MIT Press.
Smith, A. (1963). *An inquiry into the nature and causes of the wealth of nations*. New York: Wiley.
Smith, J., & Szathmáry, E. (1997). *The major transitions in evolution*. London: Oxford University Press.
Stewart, J. P. (1964). Jacobellis vs Ohio, 378 u.s. 184, Jacobellis v. Ohio. Appeal from the supreme court of Ohio. No. 11.
Strogatz, S., & Henry, S. (2000). *Nonlinear dynamics and chaos: with applications to physics, biology, chemistry, and engineering*. New York: Westview Press.
Tautz, D., Trick, M., & Dover, G. (1986). Cryptic simplicity in DNA is a major source of genetic variation.
Teisman, G. *Publiek Management op de Grens van Chaos en Orde: Over Leidinggeven en Organiseren in Complexiteit*. Den Haag, SDU Uitgevers bv.
Tesfatsion, L. (2007). Agents come to bits: Towards a constructive comprehensive taxonomy of economic entities. *Journal of Economic Behavior & Organization*, *63*(2), 333–346.
Thompson, J. (1994). *The coevolutionary process*. Chicago: University of Chicago Press.
Ulrich, W. (1988). C. west churchman-75 years. *Systemic Practice and Action Research*, *1*(4), 341–350.
van den Muijzenberg, E. (1980). *A history of greenhouses*. Institute for Agricultural Engineering.
van der Lei, T. E., Bekebrede, G., & Nikolic, I. (2009). Critical infrastructures: A review from a complex systems perspective. *International Journal of Critical Infrastructures*, *5*(4).
Von Bertalanffy, L. (1972). The history and status of general systems theory. *The Academy of Management Journal*, *15*(4), 407–426.
Von Neumann, J., & Burks, A. (1966). *Theory of self-reproducing automata*. Urbana: University of Illinois Press.
Waldorp, M. (1992). *Complexity: the emerging science at the edge of order and chaos*. New York: Simon and Schuster.
Weber, B., & Depew, D. (2003). *Evolution and learning: the Baldwin effect reconsidered*. Cambridge: MIT Press.
Whitehead, A. N. (1911). *An introduction to mathematics*. Williams and Norgate.
Wilds, R., Kauffman, S., & Glass, L. (2008). Evolution of complex dynamics. *Chaos: An Interdisciplinary Journal of Nonlinear Science*, *18*, 033109.
Williamson, W. O. (1987). *The economic institutions of capitalism*. New York: Free Press.
Wooldridge, M., & Jennings, N. (1995). Intelligent agents—theory and practice. *Knowledge Engineering Review*, *10*(2), 115–152.

Chapter 3
Practice

I. Nikolic, K.H. van Dam, and J. Kasmire

Abstract Having explored the theory of complex adaptive systems and agent-based modelling, resulting in theoretical foundations of important concepts such as socio-technical system, complexity, and agent, this theory can now be put into practice. This chapter introduces ten steps for creating an agent-based model of a socio-technical system. From the early and inherently ambiguous stages of agreeing what exactly we are interested in, through formalisation and implementation all the way to setting up and presentating detailed experimental analysis, detailed instructions for each step are provided. Examples drawn from published models and recommendations for procedures and tools are given, allowing the reader to start developing their own models.

3.1 Introduction

Like the practice of building agent-based models, this chapter starts with problem formulation using the generative science approach to look for a macroscopic regularity or pattern in the real world in order to identify and describe the problem at hand. After identification, the problem is clarified by establishing the system boundaries and the relevant level of aggregation, before describing the system in terms of the required "agents" and properties, as well as their environment and its properties. Only after specifying and formalising what "is" in the system we can move from a description of the problem to a conceptual model by describing what "happens" in the system. We do this by specifying the states and behaviours of the agents. The conceptual model lays out a narrative of who does what and when by describing the order of agent's actions and state changes. When the narrative is satisfactorily clear, modellers agree on how to translate that narrative into pseudo-code so as to make the order of events, agent behaviours and state changes and any other necessary distinctions as accurate and unambiguous as possible, as an intermediate stage before using a formal programming language to implement the simulation model.

I. Nikolic (✉)
http://www.igornikolic.com
e-mail: i.nikolic@tudelft.nl

K.H. van Dam et al. (eds.), *Agent-Based Modelling of Socio-Technical Systems*, Agent-Based Social Systems 9,
DOI 10.1007/978-94-007-4933-7_3, © Springer Science+Business Media Dordrecht 2013

After discussing how to identify the problem, how to develop a suitable conceptual model, and how to implement it, we move to verification, and experimental setup. Attention is given to matters of how to identify relevant metrics, scenarios, parameter sweeps, as well as an exploration of the data analysis and visualisation techniques for the large amounts of data from the simulations. Finally, the chapter ends by considering model validation issues, model outcomes, and the ways the model can be used to answer the original research questions. Guidelines are also given to avoid common pitfalls in interpreting the results, applying the model results, and supporting decision makers.

This chapter will walk the reader through the following ten practical steps for creating and using an agent-based model of a socio-technical system:

Step 1: **Problem formulation and actor identification** in Sect. 3.2.
Step 2: **System identification and decomposition** in Sect. 3.3.
Step 3: **Concept formalisation** in Sect. 3.4.
Step 4: **Model formalisation** in Sect. 3.5.
Step 5: **Software implementation** in Sect. 3.6.
Step 6: **Model verification** in Sect. 3.7.
Step 7: **Experimentation** in Sect. 3.8.
Step 8: **Data analysis** in Sect. 3.9.
Step 9: **Model validation** in Sect. 3.10.
Step 10: **Model use** in Sect. 3.11.

Each of these steps is accompanied by practical guidelines, examples drawn from published models, and recommendations for specific tools that have proved to be useful in collaborative model development. By the end of this chapter, and in conjunction with instruction manuals for the software tools of choice, readers should be prepared to start developing their own models and performing simulations. Part II of this book examines a number of case studies in different domains in order to detail how these ten steps can be applied.

3.2 Step 1: Problem Formulation and Actor Identification

Modelling usually starts with a problem that requires solving. Perhaps a research student has identified a research question that has not been well investigated yet, or has collected some complex and messy data and wants to understand what is really happening. Or a business owner, government agency, or industry group has realised that some issue has become increasingly pressing or that the current wisdom on how to deal with some situation appears to be working poorly. These problems frequently present as a lack of knowledge about a real-world system, its behaviour, or its response to a particular intervention. The successful model serves as a laboratory for *in silico* experiments, which if done properly are less difficult, costly or dangerous than *in vitro* experiments. Through these simulated systems, we can increase our insight into the possible nature of the real-world systems, how they function, and how they may change as a result of interventions or efforts to influence the system behaviour. The ultimate objective of a modelling exercise should be an insight, not necessarily "numbers" (Hamming 1962) because it cannot be stressed enough that modelling is a means to an end rather than a goal in itself. Models are a tool to improve our understanding of the dynamics of whole systems or subsystems, explore possible futures or find states to be approached or avoided by allowing experimentation with "what-if" scenarios, and Epstein (2008) lists "sixteen reasons other than prediction to build models".

Well-formulated problems are more likely to yield useful insights, so the first aim is to be clear and specific about the problem to be addressed and how we can get more information about it. As we are operating from a Generative Science perspective (Epstein 1999) and want to build a bottom-up system description, we begin by identifying the emergent patterns of interest, followed by the problem owner, other actors involved in the system, and our own role in both the system under study and in the model development process. In cases where we want to model not only an existent system, but also a hypothetical alternate system, we must identify emergent patterns for both systems, and generate a working hypothesis on why the two are different.

The above leads to a series of questions that need to be posed and answered before the start of a model development exercise:

- What is the problem?
  - What is the exact lack of insight that we are addressing?
  - What is the observed emergent pattern of interest to us?
  - Is there is a desired emergent pattern, and if so, how is it different from the observed emergent pattern?
  - What is the initial hypothesis on how the emergent patterns emerge, or, why do the observed and desired emergent patterns differ?
- Whose problem are we addressing?
- Which other actors are involved?
- What is our role?

Regarding the last question, it is important to mention that this chapter is written from the perspective of a modeller, and it gives practical guidance in building models to support the problem owner, taking into account the position of the stakeholders and incorporating the knowledge from domain experts.

3.2.1 Step 1 Example

The output of the problem formulation and actor identification step is an answer to the above questions, specifying the departure point for the modelling process. An example drawn from an agent-based model built using these 10 steps is provided below.

Example: Westland Greenhouse Cluster The effect of an innovation on society (e.g. increased sustainability) is a function of its diffusion, or the number people that adopt the innovation rather than the number of potential adopters that could make use of it. This example, whose background was introduced in Chap. 2, addresses the diffusion of greenhouse horticultural technology innovations throughout a group of greenhouse growers in the Westland (NL) greenhouse area, where growers often use state-of-the-art technologies to produce fruits, vegetables and flowers for the local and international market (Kasmire et al. 2011).

What Is the Problem The main lack of insight addressed is the poor understanding of the factors affecting technology diffusion between greenhouses. The observed emergent pattern is the unequal distribution of technologies among a group of greenhouse farmers, with many farmers owning sub-optimal technologies, despite the availability of better technologies and awareness of the advantages. Interviews with greenhouse farmers and growers associations revealed that farmers generally rely on intuition and a gut feeling when making decisions and that these farmers respect and value the opinions of other farmers, but are not inclined to trust information from journals, government sources or industry reports. Therefore, the flow of information and opinions among greenhouse farmers and their subsequent technology purchasing decisions are to be modelled, replicating not only the currently observed uneven distribution, but also exploring what conditions might lead to increased adoption of better performing greenhouse technologies.

Initial Hypothesis The initial hypothesis for the difference between observed behaviour and desired behaviour is that both technologies (e.g. different combined heat and power systems) and greenhouses (e.g. size, crop) are heterogeneous, offering a great number of possible combinations of technology and greenhouse types, while the trusted sources of information (e.g. the opinions of other farmers) is patchy and slow moving. Some technologies perform better with respect to profits, so the profitable technologies should be used more often. However, due to imperfect and slow information flow, and the fact that technologies cannot be replaced immediately, not all farmers will use the technologies that are the best for them at all times.

Whose Problem Are We Addressing? The main problem owner is the Organisation for Greenhouse Technology Development, a growers society promoting technological development in the greenhouse sector. While actively promoting new technology adoption by their members, they observe that many members do not adopt the most efficient and cost effective technology currently available for various reasons.

Other Actors Other actors involved include the greenhouse growers themselves, technology development companies and grower organisations.

Our Role Our role is that of the modeller, extracting knowledge from stakeholders and domain experts, developing the model and providing the insights to the stakeholders. One of our most important tasks is showing under which assumptions the model results will hold.

3.3 Step 2: System Identification and Decomposition

After the problem has been formulated, the next step towards building a model of a system is deciding on the system composition and boundaries. We want to identify the internal structure of the system under analysis to permit a generativist description. This means that the system is considered as a collection of agents that exist, have existed or may exist in the future, and all of their interactions over time giving rise to the system's emergent patterns.

Having identified the problem owner and relevant system actors in Step 1, the system identification and decomposition process requires modellers to "get social" in order to obtain the necessary information. This information could be obtained through brainstorming sessions with the problem owner and reasonable relevant entities, through surveys, literature reviews, or other social science data collection methods, or through careful observation of the behaviour of actors involved. Systems of the complexity described in this book require a great deal of intense work to fully decompose. This might be months of surveys and interviews or several full days of interactive brainstorming workshops with stakeholders and the problem owner. Importantly, the systems under study are so large and complex that they can only be interpreted from a limited viewpoint, meaning that all information gathered will contain many simplifications and assumptions. Even the modeller will have some assumptions that will be impossible to totally avoid. It is important to remain aware of the limitations of any interpretation of a large and complex system, and to keep track of which assumptions were made and why. In later stages these assumptions will be challenged and if they do not hold, then the system decomposition has to be adjusted accordingly.

The decomposition process starts with an *inventory* phase followed by a *structuring* phase before it can be *formalised* in Step 3.

3.3.1 Inventory

As part of the decomposition of a socio-technical system, the inventory phase seeks to explicitly identify the physical and social entities of the system and the links between them. This phase elaborates on the list of actors found in the problem formulation step, usually through consultation and brainstorm sessions with the stakeholders, and focuses on the following:

1. Collect data, surveys, interviews and brainstorm sessions with domain experts, stakeholders, and relevant actors. Let them talk about the system in relation with the problem. Make an inventory of all relevant elements, issues, factors, worries and explanations that come up.
2. Choose a time frame that is important. Select the longest and shortest time periods that are relevant for the system.
3. Specifically identify:
   a. Relevant concepts
   b. Actors or objects
   c. Relevant behaviours
   d. Interactions or flows (continuous or discrete)
   e. States or properties.

Modellers need not spend too much time trying to perfect the inventory at this point because it will be iteratively improved during the structuring phase, and will likely be revisited on occasion throughout the course of the modelling exercise. Any generative theorist must continually reexamine every agent, behaviour and property to decide whether it is worth capturing in the model. The real system is composed of entities composed of other entities and actors that perform all sorts of tasks, but modellers have to ask themselves whether it will help solve the problem to include the specifics. For example, a company has several levels of management, which change with some regularity, but modelling the company as a single agent might be a better way to look at how they interact with global markets. Likewise, we could model the fact that actors put a stamp on an envelope before sending out a contract, but it doesn't seem to bring us any closer to understanding why contracts are created or not. Throughout the entire modelling process, details of the model may look more or less important to solving the original question, and consequently they may be added or removed at some later stage.

Inevitably, modellers have to make some important decisions about what they believe really contributes to the problem at hand. Unfortunately, there are no hard rules for what should or should not be included, so the concepts, actors, objects, behaviours, interactions and states from the inventory that emerge from the data collection or discussions with the stakeholders should be made explicit (recall the discussion on Ashby's Law of Requisite Variety and the role of simplification in Sect. 2.6.1). Only explicitly stated ideas and assumptions can be compared to later incarnations of the model or to theoretical models from literature.

3.3.2 Structuring

The structuring phase is the core of the system decomposition and consists of three subtasks:

1. Structuring of agents and interactions;
2. Iteration; and finally
3. Identifying the external world.

3.3.2.1 Structuring of Agents and Interactions

The elements identified in the inventory phase (Sect. 3.3.1) are first grouped in two classes: agents and interactions. Agents are the basic units of the model, representing one or more actors in the system. They are recognised by their boundaries, their states, behaviours and ability to interact. For example, a basic model element may be a company. The boundaries are determined by the company's scope of control, and may include other objects or agents within them, such as a production facility. The company can make purchasing and pricing decisions which constitute its behaviour, and it is able to interact with other companies through purchases and delivery of products.

Interactions are the ways through which agents affect each other. Interactions may be one short term or long lasting, immediate or delayed. For example, if agents represent companies then the act of purchasing resources from another agents is a short term, immediate interaction. If an agent represents a government, declaring the intention to impose a tax on trade between company agents would be a delayed, long lasting interaction.

The structuring will then proceed as follows:

1. Given the results of the inventory, consider actors and objects with useful boundaries (physical, organisational and functional). Entities that are capable of independent decision making will be the agents, all others are considered to be objects. Agents can contain or interact with objects (such as companies owning facilities, or a postman processing a letter).
2. Within entities defined above identify properties that describe and specify the agents. These are states.
3. Within entities defined above, search for interactions with agents or object outside, both incoming and outgoing. These are interactions.
4. Within entities defined above, identify state changes that are caused by interactions or by other state changes, and state changes that lead to interactions or other state changes. These are behaviours.
5. Note which agents, interactions and behaviours are dynamic and which are static and at what time frame.
6. If necessary, organise agents hierarchically by ordering them in nested way, as a box within a box (e.g. departments within a company).

3.3.2.2 Iteration

The iteration provides an opportunity to reevaluate the concepts from the inventory. Concepts that were identified, but deemed to be irrelevant or tangential can be thrown out, while concepts missing from the initial brainstorming list but that have popped up as a result of the inventory phase can be added.

1. Check if there are any concepts (i.e. agents, objects, states, interactions and behaviours) identified in the inventory phase but that have not been used in the structuring phase.

   a. For these identified but not used concepts, decide whether they should be either used or eliminated.
   b. Simplify the concepts as much as possible, while attempting to retain all relevant aspects.[1]
2. Check if there are any agents with neither ingoing nor outgoing interactions.
   a. For these identified agents with missing interactions, decide whether they should have interactions added or whether the agent should be eliminated.
   b. Note that not all agents necessarily have both in and outgoing interactions as it depends on the system being modelled.
3. Check if there are any interactions described in the inventory, but that are not connected to any agents.
   a. Decide if these interactions are unconnected because there is a missing agent, or if the interaction is superfluous.
   b. Interactions cannot be unconnected, although agents may not have in and out going connections, so be sure to check all of the connections from both sides.

3.3.2.3 Environment

The structuring phase concludes by considering the world outside the agents. All the system components that cannot be influenced by the other subcomponents can be grouped together to form the external world or the *environment*. These are often referred to as exogenous variables. What will be in the environment is highly dependent on the problem that is being posed and the scope of the model. As a rule of thumb everything that is not influenced by the agents within the system but that does affect them is a part of the environment. Extremely slow processes (relative to the chosen time frame) should also be considered to be a part of the environment.

For example, things such as weather, global prices for goods in the case where we are modelling small local markets, government policies in situations when the agents cannot affect such policies, or technological innovation tend to be the environment. Predetermined features, such as a given social network or existing institutions can also be considered to be part of the environment.

Please note that the environment does not have to be static. It may have a complex dynamic, such as changes in the world market prices or a sudden introduction of a tax or regulation, and parts of the environment might even be dynamically determined by an external simulation, for example a load-flow calculation of the electricity grid, or input-output economic models. The environmental variables should be treated as all other model elements in the subsequent modelling tasks.

[1] It is impossible to give a clear and exact advice here, as each individual situation will require different levels of relevant detail. That is the art of modelling!

3.3.3 Step 2 Example

The final output of this step, reached after several iterations of the substeps, comprises:

1. a list of agents, their properties, their actions and their interactions;
2. a specification of the actions performed by the agents, used to affect their own state or for communication between the actors; and
3. a description of the environment of the elements.

The output of these three steps in the early stages of the greenhouse model can be found below.

Inventory—from greenhouse farmers surveys/interview, guided by a review of technology diffusion theory literature:

- Concepts are preferences for technologies, competition, trust for some information sources but not others.
- Actors are greenhouse growers, technology company salesmen, grower associations, greenhouse operation suppliers.
- Behaviours are growing and selling crops, talking to neighbours, making decisions and purchasing new technologies, going bankrupt if not profitable.
- Interactions are information sharing, buying and selling in markets.
- States/properties are bank balance, characteristics of the company (size and crop type), technologies owned.

The identification of agents after the structuring phase can be found below.

Agents are companies and they:

- grow stuff
- sell products
- have a certain amount of money
- make surface area dependent profit per year
- own technologies
- know about some technologies (including the ones they own)
- have a certain satisfaction about the technologies they currently own
- have a set of opinions about the performance of all known technologies
- have a stubbornness, describing how much they value their own opinion with respect to others
- have an opinion change rate, describing how much new opinions weigh against old opinions
- have a probability to try unknwon technologies
- have company specifications, which are
  - their surface area
  - their company type (either vegetable or flower, each of which has different
    - base costs (regardless of any technology or operation)
    - surface area dependent operating costs (regardless of use of technology)
    - base production (amount of crop produced regardless of the technology).

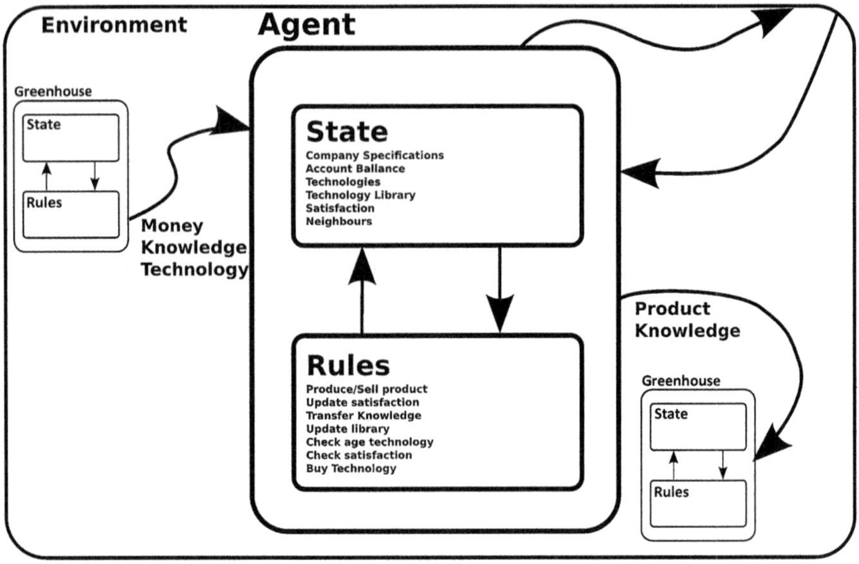

Fig. 3.1 Greenhouse example schematic model layout

The simple environment for this model can be found below.

Environment

- buys all crops produced by greenhouse growers
- sells all technologies for greenhouse growers to purchase
- absorbs the operating costs of greenhouse operation
- randomly distributes the size of the greenhouse companies
- creates the social network links between greenhouse farmers.

For a schematic overview of the model layout, please see Fig. 3.1

3.4 Step 3: Concept Formalisation

Having identified the system and the agents, as well as their states, relationships, behaviours and interactions, our next step is to formalise these concepts. The main reason for formalisation is that even though the identified concepts may seem well-defined to the stakeholders, they may be far more context dependent or specific than the stakeholders realise and computers are ill-equipped to deal with ambiguity and context dependency. For example, the concept "contract" may be interpreted as a written documents fully specifying all of the conditions, or it may be an oral agreement. A contract may contain several numbers, such as the price, the quantity, the delivery date, etc., but a computer will not know which number has which meaning until we make it specific. It is this ambiguity that makes computers unable to process a "contract", or any other concept that came out of the previous step, without further formalisation. After the inventory and structuring phases of the system decomposition have been completed, concepts are in a natural language which is not interpretable by a computer. Your model of the world needs to be made explicit, formal, and computer-understandable (in addition to being human-understandable). This formalisation also makes it possible for the system description to be generalised beyond one specific domain.

We will discuss two options for formalising the identified concepts. The first is a non-structured list of software data structures and the second a formal ontology. The advantage of the former is the ease of creation, while the latter has the advantage of reusability for future modelling instantiations, but it depends on the complexity of the model which is the best approach.

3.4.1 Software Data Structures

Starting with the list of concepts identified in Step 2, we need to convert them into computer understandable analogues. Computer languages can deal with a fairly small set of basic concepts, or so-called primitive types. For example, a non-exhaustive list of these primitive types may be:

- Numbers (of various formats, e.g. integer ($\mathbb{N}$) or floating point ($\mathbb{R}$))
- Strings (a string of characters, essentially text)
- Booleans (representing truth values in logic, which can have a value of TRUE or FALSE)
- Objects (elements containing both data and functions)
- Classes (types of objects)
- Lists and tables (of various types, containing any of the above).

The exact formalisation is dependent on the programming language used. Higher level languages may offer fairly rich and complex primitive types, such as tuples, hash tables and dictionaries (e.g. in *Python*) or linked lists (e.g. in *Lisp*) and more advanced data structures can be created out of these primitive types. For example, an agent's name will usually be defined as a *String* of characters, and the agent's balance can be formalised by stating that there is a variable of the *Integer* type, called balance. These variables can then be given a specific value for a specific agent. It is important to maintain *unit consistency* throughout the model (e.g. use Euro for money everywhere in the model, instead of mixing it with prices in for example US dollar or Pound Sterling) when formalising concepts this way. Software primitive types cannot explicitly describe unit types, (i.e. if you have declared that the balance is an Integer then it is just a number without a specific unit) and you have to implement any checks for unit conversion errors yourself.

While this approach to formalisation begins easily enough, it can pose several challenges. Complex concepts may require the construction of tables or lists, or even lists of lists, which quickly become unwieldy to work with or keep track of. Furthermore, ill-defined concepts may be difficult to represent. For example, is the concept 'trust' a number from 0 to 100, or is it a elaborate hierarchically structured construct consisting of many other objects and their relationships? Such choices have to be made when formalising a conceptualisation, but software declarations may tempt the developer to make ad-hoc decisions based on ease of implementation rather than the nature of the concept that is being addressed.

3.4.2 Ontology

On the other hand, the development of a so called "ontology" forces one to think carefully about how to formalise the concepts that have been identified, and the relationships between them, without focusing on the software implementation only. In fact, an *ontology* can be defined as being "*a formalisation of a conceptualisation*" (Gruber 1993). In the creation of an ontology "what" is in the model is formally encoded, including the objects, concepts, and other entities that are assumed to exist within the system boundaries and the relationships that exists between them. This is exactly what follows from the system decomposition process. Noy and McGuinness (2001) provide a solid introduction in the development of new ontologies and the knowledge-engineering discipline. However, it should be stressed that there "is no single correct ontology-design methodology" (Noy and McGuinness 2001).

Once a concept has been defined, it can in turn be used to define other concepts. For example, after a developer has formalised the concept "contract" (i.e. made formal what a contract *is*), it can afterwards be used to describe the concept of a company who may *have* one or more contracts. The relationships between the concepts in the ontology are thus made explicit in a hierarchical fashion. When a formal ontology already exists (for example the one used in Part II of this book and described in Sect. 4.2), the formalisation process can refine this generic ontology by creating new abstract classes applicable for the current case and by adding properties to already existing classes, or re-using concepts from one ontology into another.

When the ontology has been defined, the model specification can be finished by creating concrete instances of the abstract classes from the ontology. An instance is a single identifiable object within the limits of the scope of the model and in this view a class can be considered as a "generalisation" of a number of instances. Because all concepts are formalised in the ontology, it is possible to perform checks on the input data to ensure it is consistent with the specification and that any requirements for the data are met.

There are several tools available to assist the developer in building, maintaining and instantiating ontologies, such as OilEd, OntoEdit, and Protégé. These tools use a Graphical User Interface (GUI) for entering class definitions and some tools provide a GUI for knowledge acquisition using user-defined forms for entering information about instances. In this book Protégé is used, as it is a user-friendly tool which is available freely as open-source and is supported by a large community of users in a wide range of application domains. Protégé supports many different standard languages for storing ontologies, including W3C's XML-based Web Ontology

Language[2] (OWL) or Resource Description Framework Schema (RDFS) (Gennari et al. 2003) as well as Frames (Wang et al. 2006).

When applied to agent-based modelling, the ontology is used by the agents to exchange messages and information, as well as for the storage of data about the elements in the system. This way the agents have a shared model of the world (Aldea et al. 2004). When a model is implemented, the ontology forms the basis of the class structure for object-oriented software implementation: The 'is a' relationship is coded as the subclass relationship in class descriptions and the 'has a' provides information on the properties of the class and the possible values. Software data structures discussed above are then the *output* of this process.[3] Ontologies are not only useful for communication between agents or in the model development, but also for sharing knowledge between modellers, domain experts and users. No misunderstanding should be possible so a shared language is needed, and ontologies are meant to be not only computer-understandable, but also accessible to human users.

Ontology design is not a straight-forward task and it is quite likely that early designs will require a major overhaul based on new insights into how to best formalise the concepts that play a role in the system. Often the process is a group effort too, with several people involved sharing their views on how to transform the concepts that are the outcome of Step 2 into a non-ambiguous definition. The result of this effort is a strong foundation for the development of the agent-based model.

3.4.3 Step 3 Example

The outcome of the concept formalisation step is a precise description of the concepts that play a role in the system that we are modelling, including the agents, their states and properties.

Software Data Structures First a conceptualisation in software data structures is presented. In this example we show the formalisation of the Company agent and the Technology object:

Companies have:

- money: integer.
- profit per year: integer.
- technologies: a list.
- satisfaction: integer $>= -1$ and $<= 1$.
- technology libraries: list of lists.
- opinion library: list of lists of integers $>= -1$ and $<= 1$.

---

[2]See http://www.w3.org/TR/owl-guide/.

[3]Ontology development tools such as Protégé can even automatically generate Java classes based on the ontology definitions.

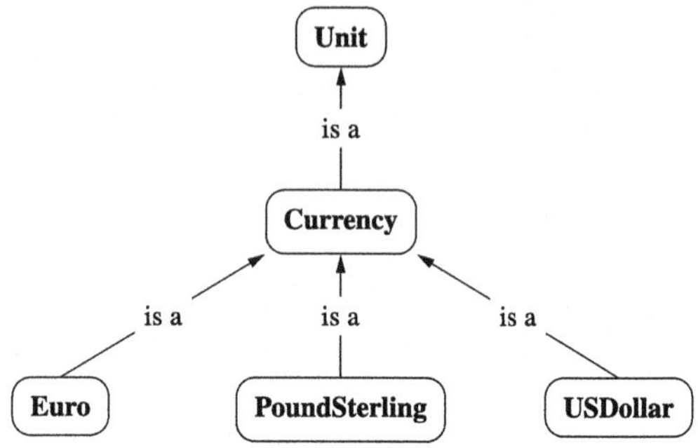

Fig. 3.2 Example ontology of Currencies as a Unit

- stubbornness: integer >= 0 and <= 1.
- opinion change rate: integer >= 0 and <= 1.
- company specifications:
  - surface: integer > 0.
  - company type: integer. Each company type has different:
    - base costs: integer >= 0.
    - operating costs: integer = 0.
    - base production: integer >= 0.

Technologies have:

- category: integer.
- price: integer >= 0.
- cost: integer. The company has a list of these values.
- performance: integer. The company has a list of these values.
- lifespan: integer >= 0.
- age: integer >= 0 and <= lifespan.

Ontology Alternatively, we will consider a possible formalisation of the concepts for the greenhouse model in an ontology. Starting from the conceptualisation of the model from Sect. 3.3.3, we can see that there are various properties related to the economics of the companies and the technologies, including profits, costs and prices. To make these figures meaningful they need a unit of currency in which the specified amount is calculated. Start the ontology development by saying that there *is* a thing called a Unit with the subclass Currency (see Fig. 3.2). Stating that Currency 'is-a' Unit and a subclass to Currency called Euro exists, which 'is-a' Currency, means that Euro is also a Unit through class inheritence. We can continue by saying the USDollar 'is-a' Currency and that PoundSterling 'is-a' Currency, etc.

By formalising these concepts, we have added some meaning to these words, which can be used in the rest of the formalisation. We can now say that a Price 'has-a' numerical value, using the base class Integer, and that it 'has-a' unit that must be some subclass of Currency. As currency units, the Euro, US Dollar and Pound Sterling are valid choices to describe a price. Furthermore, this definition can also

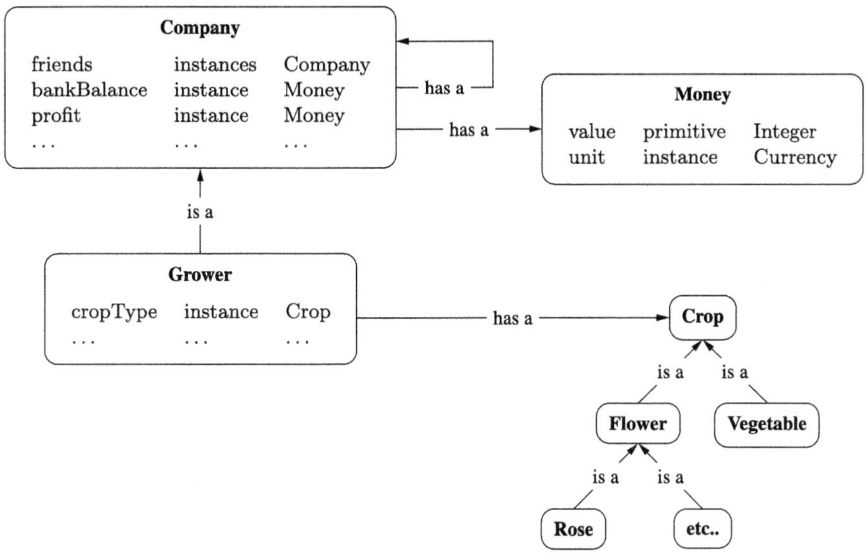

Fig. 3.3 Segment of the example ontology for Growers, re-using other concepts such as Currency as defined in Fig. 3.2

be re-used to describe profits, account balance, or any other economic properties in the model, thereby ensuring that the units used for different properties are consistent in meaning.[4]

Similarly with higher level concepts, we can define that a Grower 'is-a' Company and that each Company 'has-a' list of other companies it knows (referring to its own "class" for allowed types), and that a Grower 'has-a' crop type which is also classified hierarchically. See Fig. 3.3 for an example formalisation. Using such a structure, a Grower agent could reason about the world. For example, it knows which "flower growers" are in its network using the meaning that is tied in the class structure of the ontology: if one of its friends grows roses and another grows tulips, the agent can deduce that they both grow flowers. Thus the words in the conceptualisation start to "make sense" for the computer model too.

In the rest of this chapter, however, we opted to keep the model specification simpler and to not elaborate on the development of an advanced case-specific ontology. We will thus use the data structure presented above in the model formalisation of Step 4 and software implementation in Step 5, but we hope that the advantages of using an ontology for large and complex models, especially when multiple parties are involved, will become perfectly clear in Part II of this book where each case study employs an ontology to make model development easier, re-useable and better manageable.

[4]The consistency would be particularly important if the definition contains ambiguous terms such as "dollar" which could mean US dollar in one case, but maybe Hong Kong dollar or Singapore dollar in another. That could lead to errors in calculations unless this is clearly specified.

3.5 Step 4: Model Formalisation

Once we have identified *what* and *who* is in the model, we need to establish who does what and when. As we are building generative models, we need a "story" of how agents act and interact, thereby giving rise to emergent patterns that we are interested in.

It is not unusual for the modeller to discover that some aspects are missing as the description of the system becomes more formal. As we delve into the specification of details, we are more likely to stumble upon aspects of model behaviour and agent interactions that were not considered before. Such iterations are a natural part of the modelling process, and are relatively "cheap" in terms of effort at this stage. Skipping this step, only to discover problems during model implementation, requires much more effort and potentially rewriting code to make corrections.

Model formalisation consists of two tasks: creation of a *model narrative*, and the expression of this narrative in *pseudo-code*.

3.5.1 Developing a Model Narrative

A dictionary definition of a *narrative* is "an account of a series of events, facts, etc., given in order and with the establishing connection between them; a narration, a story, an account" (Oxford English Dictionary). A narrative is an informal description of the generative theory of the system under study, leading to emergent patterns we are interested in exploring. For an agent-based model, the behaviour of each of the agents can be captured in a story which explains *which agent does what with whom and when*. The narrative, which encompasses the individual stories of all the agent types, can be very simple or very complex. Essentially, the narrative must finish the sentence "The agent wakes up, gets a cup of coffee and then ... " for each of the different types of agents. For example, an agent wakes up, gets a coffee and chooses the cheapest contract out of the contracts offered, accepts that contract, transfers the money and receives the goods. The story can be written in almost any format, including as a list of bullet points, a short text, a flow chart, a series of graphical representations, or as a UML action sequence diagram, among other options. The model narrative must be as complete as possible because it will be useful in discussions with the domain experts, problem owners, and other modellers.

An important aspect of creating a model narrative is considering the assumptions of parallel action and its sequential implementation in a program, see Sect. 2.7.3. The discrepancies between assumptions of parallel action and the reality of randomised sequential action can be subtle. Programmers should be careful to design the model narrative so that agents cannot access information from that should only be available after the tick and to prevent first-mover advantage. Also consider the implication of active/proactive distinction between agents, see Sect. 2.7.2.2, to avoid recursion errors, in which agents call each other indefinitely. Every potentially recursive situation needs to have explicit *exit* conditions to avoid the recursive trap.

A simple exit condition could be to specify the depth of recursion or a specific and attainable condition at which point, the loop is stopped. Good programmers might use both, so that agents do not offer and reject contracts infinitely, but accept a contract after the offer is no more that 1 % different than the last contract, or when 10 contracts have already been rejected, for example.

Once we have an informal narrative, the next action is to think about a more precise algorithmic representation of that behaviour.

3.5.2 Pseudo-code

Pseudo-code is a description of an algorithm written in human-readable form, providing an insight into the structure of the algorithm while omitting computer specific details, such as variable declarations, system-specific code and subroutines. It uses mathematical and logical descriptions of what and how agents are supposed to behave to combine the model narrative and the formalised concepts resulting from Step 3.

Pseudo-code bridges the gap between an informal model narrative and specific computer code. Writing the model out in pseudo-code forces us to think about a logical mode layout, without distraction from software implementation details. This is especially useful when you are inexperienced in computer programming, but it should be stressed that even experienced programmers will find writing pseudo-code a beneficial way to separate *what* to say and *how* to say it, as well as aiding in software design and documentation or for tracing bugs.

3.5.2.1 Elements of Pseudo-code

There are several types of operations that are useful when creating agent pseudo-code. A non-exhaustive list of useful operations is given below:

- Computation and assignment
- Iterations and loops
- Conditions
- Input/output operations.

These operations to be used in pseudo-code will be discussed briefly below. Readers interested in a in-depth discussion of pseudo-code should consult McConnell and Books (1993).

Computation and Assignment When an agent needs to know which of the offered contracts is the cheapest, we would express that in pseudo-code by using words such as generate, compute, process, calculate, set, reset, increment, get, add, sum, multiply, print, display, input, output, edit, test, etc. Mathematical symbols and arithmetic expressions can be used too. Algorithm 3.1 presents several assignment pseudo-code examples.

Algorithm 3.1 Examples of assignment pseudo-code

```
set the value of "variable" to: "arithmetic expression"
"variable" equals "expression"
"variable" = "expression"
```

Iterations and Loops When an agent must do the same action several times per timestep, such as offer each of his neighbours a contract, then this repeated actions can be formalised as iterations or loops. They can often be combined with conditional statements, so that the agent will keep repeating the action until some other condition is met. For example, rather than offer all of his neighbours a contract, the agent can offer contracts until some fixed number of contracts are accepted, or until some target is reached by accumulating the value or volume of product in accepted contracts. For clarity, it is important to "scope" the loops, which means to clearly identify what is in the loop or not.

Keywords associated with loops and iterations are do While ... End while; Do until ... End until; Case... End case; Call ... with (parameters); Call; Return ...; Return; When. An example of loop pseudo-code is presented in Algorithm 3.2.

Algorithm 3.2 Examples of iteration and loop pseudo-code

```
do while "condition"
   statement 1
   statement 2
end while

for <variable> from <first value> to <last value> by <
    step>
   statement 1
   statement 2
end for
```

Conditions At the heart of every agent's decision making lie conditional statements describing how the agent reacts to a particular state (or change of state), so that agents, for example, could decide to accept a contract if the value of the contract is equal to the value of the last contract accepted. The conditional statements can be as complicated or as simple as necessary, depending on how the conceptual model describes their actions.

Conditional statements have the form of if. . . then. . . else structures. For clarity, it is important to scope the if statements as well. Examples are presented in Algorithm 3.3.

Algorithm 3.3 Examples of conditional pseudo-code

```
if "condition"
   statement 1
   statement 2
else
   statement 3
   statement 4
end if
```

Input/Output Input/output (often abbreviated as I/O) statements are meant to indicate interaction with the user, data or other programs. Displaying variables may be useful to aid debugging, or when loading external data needed to run the simulation. Useful words here include get, store, save, print, display etc. An example of several I/O pseudo code statements is presented in Algorithm 3.4

Algorithm 3.4 Examples of input/output pseudo-code

```
get "variable"
display "variable"
save variable
```

3.5.2.2 Unified Modelling Language

An alternative to pseudo code is Unified modelling Language (UML). It is a formal language that is used to specify, construct, document and visualise object-oriented software systems (including most agent-based models). The advantage of UML is that provides a unified grammar to specify relationships between system elements and there are many different software tools available to visualise them. Their disadvantage is that the user needs to be familiar with UML, which is usually only taught to software engineering students. For those modellers that know how to use it, it may offer some advantages, such as automatic class generation. Detailed specification of the actions, however, still needs to be done in code.

3.5.3 Step 4 Example

The outcome of the model formalisation step is the model narrative and the associated pseudo code.

Model Narrative Example, Actions per Time Tick The narrative in long sentence format could be as follows: An agent (i.e. a company) gets up, gets a cup of coffee, sells the products that have grown, buys any required raw materials, pays for the operation of the technologies and then calculates the profits based on the revenue minus any costs. Then, it adds the new profit to the bank account. Finally, the agent updates the table of how satisfied it is with the technologies it currently owns by comparing profit to the profit of all neighbours. If it has more profit than them, this increases satisfaction for all technologies, but, on the other hand, if there is less profit, the agent decreases the satisfaction.

Narrative in bullet point format Companies do:

- calculate profit
  - profit = revenues − costs
  - revenues = price for their product * surface area * (base production + sum(performance))
    - price for products is dependent on the company type
    - base production is dependent on the company type
    - performance is dependent on the technologies owned and the company type
  - costs = basecost + surface area * (operating cost + sum(technology costs))
    - base cost and operating cost are company type dependent
    - technology costs are dependent on the technologies owned as well as the company type
- Add profit to account balance
- Check to see if any of the neighbours has a new and unknown technology, and if so add it to Technology Library
- Update satisfaction proportional to profit made
  - Compare profit per surface area to average profit per surface area for all the neighbouring companies
  - Normalise the difference between own profit per surface and average of neighbours by converting it to a value between −1 and 1
  - Add the normalised difference to the satisfaction tables for all of the technologies owned in that timestep
  - Check to see if any neighbours have acquired any new technologies that are unknown and add them to Technology Library.

Pseudo-code Using this narrative as a starting point and taking into account the formalised concepts, we can next translate this into pseudo-code (see Algorithm 3.5).

Algorithm 3.5 Model pseudo-code example—Company opinion update

```
All companies
 calculate-profits
   set profit (ProductPrice * SurfaceArea * (BaseProduction + sum(TechologyProductionEffect))) -
       (BaseCost + SurfaceArea * (OperatingCost + sum(TechCosts)))
   set AccountBalance AccountBalance + profit
   set ProfitPerSurfArea profit / SurfaceArea

 update-satisfaction
   for each neighbour in AllMyneighbours
        ask for ProfitPerSurfArea
   calculate ProfitPerSurfArea - (sum(ProfitPerSurfArea of AllMyNeighbours) / number of companies in AllMyNeighbours)
   normalise the difference by converting to a value between -1 and 1
   save normalised difference in 'satisfaction'

 update-TechLibrary
   for each neighbour in AllMyNeighbours
        Ask what are their current technologies
   end for
   for each Technology, compare to TechnologyLibrary
        if not in TechnologyLibrary, add to
           TechnologyLibrary
   end for
```

3.6 Step 5: Software Implementation

Once we have created the model narrative and formulated the actions in pseudo-code, we must implement the model in an appropriate modelling or programming environment. But deciding on the appropriate environment is no trivial task. We have found several best-practices to help during the implementation that all new modellers should consider. This section discusses all the issues of software implementation, starting with the choice of modelling environment.

3.6.1 Modelling Environment

The modelling environment is the programming language or modelling suite that provides the software infrastructure for programming the agents, their states and behaviour, their interactions and the environment. It also offers necessary and useful support code, such as a scheduler, graph plotting, statistics collection, experiment setups, etc. There are many agent-based modelling platforms and modelling environments, each with strengths and weaknesses. Nikolai and Madey (2009) provide a good overview of the available tools. Below we discuss three tools we have used most extensively to develop agent-based models of socio-technical systems: NetLogo, Repast and custom code.

3.6.1.1 NetLogo

Netlogo[5] is a free program that allows easy implementation of agent-based models. It also has a low barrier to entry, meaning that new users should not find it to hard start using it from day one, and an extensive online community for support when questions inevitably arise. It is often used for teaching due to fairly simple programming syntax and a large number of example models. However, the simplicity of the programing language can limit the complexity of models created, so it may be necessary to interact with NetLogo through direct Java language calls for more detailed modelling points. NetLogo is especially useful for developing quick proof-of-concept models, rapid prototyping or when demonstrating the agent paradigm.

3.6.1.2 Repast

The Recursive Porous Agent Simulation Toolkit (Repast) (Nikolai and Madey 2009; North et al. 2006) is another open source and free tool.[6] Because it is much more advanced than NetLogo, Repast is completely flexible in the way agents are described and allows more control over aspects of a simulation like scheduling. Like NetLogo, Repast boasts a large user base, so when questions arise users can seek help from fellow modellers. The latest version, Repast Symphony, offers useful functions like advanced user interfaces, 2- and 3-dimensional displays, integration with Geographical Information System (GIS) data, and others. The main disadvantage is that the increased flexibility entails a relatively more complex interaction, so is much less accessible to non-programmers, and extensions may require a working knowledge of Java. All models presented in this book (with the exception of model in Chaps. 8 and 9) were built in Repast.[7]

[5] See http://ccl.northwestern.edu/netlogo/.

[6] See http://repast.sourceforge.net/.

[7] Please note that these models are developed in RepastJ, a predecessor of the current version called Repast Symphony.

3.6.1.3 Custom Code

For more unusual and complex agent descriptions (for example work presented in Chap. 8) we often resort to building agents from scratch, directly in programming languages such as Java.[8] Writing agent-based models in such a way provides enormous flexibility, allowing any combination of computing tools and software libraries to be brought together. A wealth of libraries is available for all aspects of agent programming (including data structures and visualisation) giving you significant power and flexibility, but writing custom code also means there will be less guidance and support. Modellers will need to specify all aspects of the model (from agent descriptions, scheduling, data collection, displaying graphs, etc.) and this requires a solid understanding of advanced programming languages.

3.6.2 Programming Practices

Creating and using computer models as a part of the scientific process means performing experiments *in silico* rather than *in vivo*. Without a physical reality, *in silico* experiments demand even more stringent requirements than *in vivo* experiments. All scientific experiments need to be verifiable, testable (or rather, falsifiable) and repeatable by other researchers, in order to meet the preconditions of the peer review process. Typically, physical experiments must clearly and carefully describe the experimental setup, conditions, and the methods of data collection and information processing. If a standard setup is well enough known, as perhaps for DNA sequencing, then it would be sufficient to say that it was used, but novel experimental techniques call for great detail to allow for replicating the experiment. In the same way, computer simulated experiments need to be clear and careful to describe the software implementation, experimental setup, methods and techniques used to allow the experiment to be replicated, verified and tested.

Regardless of the actual modelling environment used, there a number of best practices that should be followed when developing models. Depending on the scale of the project, several practices should be considered. When developing a one-off model alone, the following programming practices will be sufficient:

- Revision control (also called *version control*)
- Documentation while you are implementing.

In the case when we are developing a large scale model with many people involved extra practices need to be implemented:

- Standardisation of naming conventions for variables and methods
- Divisions of tasks and responsibilities
- Bug tracking.

[8] See http://www.java.com/.

In the next sections these practices are briefly introduced and their relation to model replication, verification and testing discussed.

3.6.2.1 Version Control

Version control software supports multiple modellers or developers to share the latest version of important documents, code, data and ontologies and helps to keep track of the different versions over time. There are currently several options for version control, with and without a central repository. When a project requires a central "official" repository of documents, data, code or shared files, Subversion[9] (SVN) may be a good choice. When a central repository is not needed or available, git[10] or mercurial[11] can be used.

Version control software and centralising the storage of all relevant files on a central server make accessing, using and editing shared work not only possible but much easier than, for example, emailing the most recent version to the all members of a research team. Instead of hoping that a given version is the most recent, version control software records each modification to a file, allowing users to always have access to the most recently logged version, as well as knowing exactly when the changes were made and by whom. Furthermore, version control makes it easy to precisely specify which version of a file (such as the code used to run a simulation or an input dataset), was used. This means that any given experiment can be repeated or modified by retrieving the same code used previously, while still allowing the rest of the team to continue developing the model. All version control systems function over the Internet, so sharing models or files with other researchers is as easy as sending a URL in an email.

Version control software also keeps all earlier versions of files available, which is useful even for modellers working alone. When a problem is discovered, the record of past versions makes it easier to go back to see when a problem first occurred, or to salvage work that had deteriorated with subsequent versions. Version control also permits easy synchronisation between multiple computers (e.g. desktop and laptop), allows adding comments to different version releases and provides secure back-up of files over time. We highly recommended that all modellers start using some sort of version control when designing models, embedding the use of version control in the day-to-day workflow.

3.6.2.2 Documenting Code

Code documentation is essential and should be considered an integral part of implementing the model in software. Good code documentation improves awareness of

[9]http://subversion.tigris.org/.

[10]http://git-scm.com/.

[11]http://mercurial.selenic.com/.

what is being developed and why, as well as why it has been implemented in a certain way. Proper documentation also makes it much easier to read the code, allowing for easier model reuse and enhances communication about the model. Another advantage of documenting code is that model verification requires that peer reviewers must be able to verify that the model is performing as intended and has been correctly implemented. Ideally, the pseudo-code written in Step 4 forms a good starting point. Adequate source code documentation must at least make clear what the main data structures do, the mechanism by which they are updated or changed, and any peculiarities of the algorithms used. If the code is changed as a result of testing after implementation or other debugging, the comment blocks should also be revised to accurately reflect what the code does.

3.6.2.3 Naming Conventions

Consistent naming of agent states, actions and interactions is paramount when collaborating. Variable with names like x, y or i decreases the readability of code, and increases the likelihood of introducing errors in collaborative editing. Now that all modern programming environments providing variable and function name "auto-completion" there is no excuse for not using long and descriptive variable names (e.g. agentProfitsBeforeTaxes) whenever applicable. Shortening variable names does not save time and makes understanding code difficult, so abbreviations should be avoided unless they are commonly used and readily understood. The concepts formalised in the ontology can be used to for the variable names, again ensuring the correct meaning is given to a term.

3.6.2.4 Divisions of Tasks and Responsibilities

Teams of modellers may want to split model development between different modellers, so that, for example, each modeller programs a different agent type. This divide-and-conquer approach works best when version control, clear documentation and clear naming conventions are already established, thus greatly increasing the rate of development. However, dividing tasks does concurrently increase the need for communication between modellers. A shared ontology, providing a common language for the project, can make it easier to define the "interfaces" between code written by different programmers.

3.6.2.5 Bug Tracking

Bug tracking is a standard software development practice which aids the verifiability of models, and help external reviewers understand issues quickly. Bug tracking usually consists of a web-based system for reporting issues with the software such as bugs, enhancements, and plans for future extensions. Through such a tracking system, responsibility for reported issues is assigned to a developer, and the status of each issue can be tracked to monitor fixes and new versions. Version control works well when paired with a bug tracking system. For example, TRAC system[12] is tightly integrated with Subversion, allowing for repository management, bug tracking and even a wiki for discussion and documentation. Many other bug tracking alternatives exist.[13]

3.6.3 Step 5 Example

The main outcome of this step is a software implementation of the model formalisation, ideally residing in a version control system, with ample documentation.

NetLogo In Algorithm 3.6 we present part of the greenhouses model as implemented in NetLogo. This example code is taken from an actual operational model and contains several bad practices, such as unspecific and poorly named variables (e.g. the use of "temp" and "temp2" which do not in any way help the reader understand what these variables represent) and no documentation of the code. As a result, this code is relatively difficult to read, despite the explicit syntax of NetLogo.

3.7 Step 6: Model Verification

When we have a working model in computer code we must ask ourselves an important question: did we correctly translate the conceptual model into the model code? Computers will do exactly what we *tell* them to do, and not what we *want* them to do, which are often not the same. Therefore, we must take some time to make sure that model implementation corresponds with the model design before we can continue analysing and using the model.

Verification checks that all relevant entities and relationships from the conceptual model have been translated into the computational model correctly. Put another way, verification makes sure that the modeller has "built the *thing right*". Please note that the question "Did the modeller build the *right thing*?" concerns validation of the model and will be discussed in Sect. 3.10.

Verification is always a difficult task, particularly so when dealing with the complex software developed for agent-based modelling. Complexity arises from the high number of agents, their states, and the number of possible interactions. Furthermore, the fact that we often seek to explore emergent patterns that are not known in advance can make the verification task even more onerous. We must always be vigilant, especially when we have found surprising emergent behaviours, that we have really

[12] See http://trac.edgewall.org/.

[13] See for example http://en.wikipedia.org/wiki/Comparison_of_issue-tracking_systems.

Algorithm 3.6 Model NetLogo implementation example—Company opinion update

```
to update-OpinionLibraries
  ask companies [
    set OpinionLibrary2 OpinionLibrary
    let teller 0
    foreach TechLibrary [
      let temp ?
      foreach temp [
        let temp2 ?
        ifelse temp2 = item teller Technologies [
          set OwnOpinion satisfaction] [
          set OwnOpinion item temp2 item teller OpinionLibrary]
        let teller2 0
        let temp3 0
        foreach AllMyneighbours [
          ask ? [
            if temp2 = item teller Technologies [
              set teller2 teller2 + 1
              set temp3 temp3 + satisfaction ]]]
        ifelse teller2 > 0 [
          ifelse NoiseChance >= random-float 1 [
            set OtherOpinion random-float 2 - 1][
            set OtherOpinion temp3 / teller2]] [
          set OtherOpinion item temp2 item teller OpinionLibrary]
        let temp4 (OwnOpinion * Stubbornness + OtherOpinion * (1 - Stubbornness)) * OpinionChangeRate + item temp2 item teller OpinionLibrary * (1 - OpinionChangeRate)
        set OpinionLibrary2 replace-item teller OpinionLibrary2 (replace-item temp2 (item teller OpinionLibrary2) temp4) ]
      set teller teller + 1 ]]
  ask companies [set OpinionLibrary OpinionLibrary2]
end
```

learned something new from the model system rather than merely observing what appears to be emergent behaviour resulting from programming artifacts.

Nonetheless, there are four main parts to verifying agent-based models:

1. Recording and tracking agent behaviour, in which relevant metrics are identified and recorded.
2. Single-agent testing, in which the behaviour of a single agent is verified.
3. Interaction testing limited to minimal model, in which the interaction between agents is tested.
4. Multi-agent testing, in which the emergent behaviour of multiple agents is examined.

These phases are introduced in the following sections.

3.7.1 Recording and Tracking Agent Behaviour

To verify the model operation relevant output variables must be selected and monitored. These variables may be the same as those that will be recorded as output from the final experiments, but this is not necessarily the case. Verification will likely want to take a closer look at individual agent behaviour, as defined in the model formalisation discussed in Sect. 3.5, to be sure that the model is operating as expected at the level of the agent, in addition to correct operation at the level of the system where any emergent behaviours will most likely be visible. There are several options available record and track agent behaviour:

- Record the inputs, states and outputs of agents.
- Log the inputs, states or outputs of each of the internal processes.
- Walk through the source code using a debugger.

Firstly, one can record the inputs, states and outputs of individual agents. This records what an agent's behaviour looks like from "outside" the agent. If agents are simple, for example if they only have one decision making process, then this may be enough to deduce whether agents behave as desired or not.

Furthermore, the modeller can choose to log the input, state or output of each of the internal processes of individual agents. This allows the modeller to see all of the "thought" processes internal to an agent that are not visible from the final actions or output. If agents have complicated decision making processes, or if they make several internal calculations before taking any action, then recording all of the internal goings on should read as a "conversation" between agents which recreates the model narrative. For example, in the greenhouse growers model, every tick means that an agent must calculate its profit, compare this profit to that of the neighbours, adjust the satisfaction with the technologies accordingly, investigate whether the neighbours have any new technologies and add them to the agent's technology library, and then incorporate the opinions of neighbours about all technologies into the agent's opinion tables. Any unexpected behaviour, such as agents never buying any technologies of the type that was not owned by the agent already, could result from a problem at any of these actions (but we would not know where the problem was unless we recorded the internal states of the agent).

Next, walk through the source code using a debugger. Any integrated development environment should have a built in debugger that allows the programmer to go through the code, line by line, to set "break points". These allow the programmer to track changes, such as resetting the value of a variable, that show how the model is operating in detail. Such a debugger could be an alternative to recording the internal states of agents, but inexperienced modellers would do well to try both in order to have an informed opinion about which would be best to use based on the specifics of the model.

3.7.2 Single-Agent Testing

Because of the complex nature of agent-based models, often consisting of a large number of agents, an important, but rarely considered action during the verification step is exploring the behaviour of a *single* agent. It is often wise to implement these tests in your software platform rather than doing it manually so that they can be repeated when needed. In a technique called *unit testing* small parts of the code are tested by predefining inputs and listing expected outputs, after which the tests can be run automatically to give insight into how the code responds to these tests. Furthermore, even if your code changes over time you will always know whether it passes the tests, after which you will either have to debug your code or, in some cases, redesign the tests if they are no longer applicable. Some programmers even write tests *before* the actual code to be tested has been implemented, which helps to think about what exactly the lines of code you are about to write are supposed to do and gives a clear goal to work towards.

To properly explore the behaviour of agents, there are two fundamental tests, explained in detail below.

3.7.2.1 Theoretical Prediction and Sanity Checks

The conceptual model and model narrative make general predictions about how a typical agent will behave under normal operating inputs. In this test, we make explicit predictions of what we theoretically expect the agent to do when we provide well-defined inputs to agents. For example, if we expect our agent to accept a contract for a product it needs, then we manipulate the input so that the agent is offered such a contract and we record whether or not it accepts it. Any deviation from the theoretical prediction directly points at an implementation error. If our agent had refused the contract, we would know that the decision making is not operating correctly, that it was somehow unaware that it needed the product offered, or some other error that prevented the agent from acting as expected. The pseudo-code and actual code should be compared to identify whether there is a logical error in the pseudo-code or a practical implementation error.

If agents act on time series or have some kind of memory, then special care should be taken to make theoretical predictions about the behaviour with extremely different time signals. For example, if the agent must offer prices for products based on recent market prices for that product, what kind of prices does it offer when the market has recently been random, continuously increasing or decreasing, moving according to functions or power law distributions? Furthermore, when agents have a memory, we should check whether artifacts arise when we change the default length of an agent's memory. Does it behave as we would expect it to if it uses the last two market prices to make an offer, as well as the last ten, twenty or more market prices? Even if we do not identify problems, knowing how the agent's decision making changes as a function of memory size is useful later when interpreting model outcomes.

3.7.2.2 Breaking the Agent

If the agent behaves as expected under normal inputs, then the next task is to attempt to "break" the agent and define the edges of normal behaviour. This means testing the parameters or parameter time series with extreme values, values on the edges of the expected range, or values that may provoke errors, until the agent performs unexpected actions or behaves in very different ways after crossing an input threshold. When feeding the agent extreme input, such as very large ($+\infty$) or very small ($-\infty$) values, it might react in a completely appropriate and logical way, thus verifying the agent, or may react in a logical but surprising way, revealing that there is a threshold for input at which its behaviour will change. Alternatively, the agent might do something unexpected or break altogether.

Unexpected actions are not necessarily implementation errors, instead they only establish the limitations of the computer code. Consider that an agent receives a 0 for some parameter value which is later used as the denominator in a later calculation. Dividing by zero will indeed break the agent, but this is clearly not an error in the code. Finding such a limitation is not the end of it, however, because we need to be sure that an undesirable value could not be generated by some other agent during normal operation. The code would benefit from modification so that some agent (either the one who could generate the breaking value or the one who would break upon receiving it) can gracefully deal with it. Perhaps the agent performing the division could check for 0 before attempting to divide, the agent creating the value could check to see that it falls into the proper range before sending the value out or, ideally, both.

3.7.3 Interaction Testing in a Minimal Model

After testing a single agent, we need to test agent interaction by following the same tests as defined for single agent testing. The difference will be that we are performing the theoretical predictions, sanity checks and breaking the agent limitation tests when the model is set up to run with the absolute minimum number of agents necessary. If the model has only one agent type, the minimal model would be two agents, but if there are multiple types of agents, the minimal model should include at least one of each type. We examine whether the basic agent interactions happen correctly as laid out in the model narrative. Are the agents able to find each other and if so, do they interact as designed?

As with the single agent testing, there are two important things to check this way: whether we get the desired interaction and whether we get "undesired" or unintended interactions. It may seem repetitive to run through all the same tests when they only difference is that there are now two agents instead of one, but time spent paying attention to details at this stage will be time saved later when it comes time to analyse results.

3.7.4 Multi-agent Testing

Once we are satisfied with the minimal model verification, we should move to verifying the entire model with all agents present through the same theoretical prediction and breaking the agent tests as used before, as well as two additional tests that only apply to the full model. If the preceding tasks have been thoroughly performed, this final check should not reveal any new surprises so the theoretical predictions should be easier and more accurate. Similarly, varying the parameters in order to break the agent should be intuitive as the previous verification tests will have revealed how agents are likely to fail.

Testing for coherent behavioural patterns at this stage might reveal model behaviour that does not correspond with the issues identified during the problem identification step. Verification however is only concerned with whether the model is capturing the desired behaviour, not whether it is capturing additional behaviours as well, so full consideration of the model behaviours is better addressed in the validation stage.

The two additional tests to perform are variability testing and timeline "sanity", as discussed below.

3.7.4.1 Variability Testing

Agent-based models are often chaotic, as the order of agent interaction may constantly change, so modellers should take care to uncover any interaction-order artifacts by exploring the variability of the output corresponding to different regions of the parameter space. The best way to check for such artifacts is to perform many (e.g. 100 to 1000) repetitions of the model and examine the statistics of the outcomes, across a number of output variables. Do the standard deviations and variance make sense or do they indicate a problem? Is the skewness or kurtosis severe enough to suggest large outliers? Closer inspection will identify whether an artifact is present or whether the results can be explained by non-linear chaotic model behaviour. A model may be chaotic in one part of the outcomes, but not in other, so all parameters should be checked thoroughly.

3.7.4.2 Timeline Sanity

As a final verification of the model implementation, the timeline of the model run should be examined by performing several runs at the default or representative parameter settings. Stakeholders and modellers may identify behaviour that do not seem to make sense throughout the course of a model run. If any part of the output cannot be explained by reasoning through the model logic, implementation errors are likely to be the cause of unexpected results. Does anything unexpected happen? If so, can it be explained by following the models logic? Does it generate the required regularity identified in the problem formulation step? If not, there might be an implementation error.

3.7.5 Step 6 Example

The main outcome of this step is a series of tests with the agents and agent population, detailing what and how has been tested and presenting the outcomes. If necessary, the modeller can go back to the conceptual design (including the narrative and pseudo-code) to start tracing back the cause of any issues. The code can be fixed only after you have understood what causes the problem. This example demonstrates interaction testing in the Greenhouse model.

Model verification example—single agent and minimal model interaction testing

Single agent: Update-Satisfaction

- If all neighbours have a Profpersurf of 100000 (assuming normal profits are way below this figure), the Satisfaction of the single agent should be -1 for all the technologies currently owned. **Confirmed.**
- If all neighbours have a Profpersurf of -100000 (assuming normal profits are way above this figure), the Satisfaction of the single agent should be 1 for all the technologies currently owned. **Confirmed.**
- If all neighbours have a Profpersurf of 0, the Satisfaction of the single agent should be 1 or -1 for all the technologies currently owned, depending whether its Profpersurf is positive or negative. **Confirmed.**

Single agent: Update-opinionlibraries

- If one neighbour has an opinion of 1 for a given technology (and the Opinionchangerate is 1 and the Stubbornness is 0) the Opinionlibrary of the single agent should change from 0 on all technologies to 1 on that given technology (rest remains 0) after one tick. **Confirmed.**

- If one neighbour has an opinion of -1 for a given technology (and the Opinion-changerate is 1 and the Stubbornness is 0) the Opinionlibrary of the single agent should change from 0 on all technologies to -1 on the given technology (rest remains 0) after one tick. **Confirmed.**
- If one neighbour has an opinion of 1 for all the technologies (and the Opionion-changerate is 1 and the stubborness is 0) the Opinionlibrary of the single agent should change from 0 on all technologies to 1 on all technologies. **Confirmed.**

Minimal model: Technology and Satisfaction update

- If one neighbour has a given technology but another does not, then after the technology update code, the new technology should appear in the second agent's technology library. **Error found.** Technologies were added to the second agent's technology library that did not correspond to technologies owned by neighbours. **Fixed, revalidated, confirmed.**
- Satisfaction should be normalised to a value between -1 and 1. **Error.** The values were always appearing as either -1 or 1, but nothing in between due to a wrong variable in the code (temp2 instead of temp).[14] **Fixed, revalidated, confirmed.**

3.8 Step 7: Experimentation

Now that we have created and verified a software implementation of a model it is time to perform the experiments that may provide us with the insights into the nature of the macroscopic regularity of interest described in Step 1. We also explored whether the model should attempt to replicate some real-world regularity in order for us to understand what causes it, or rather, to create a variety of possible conditions in order to investigate what kinds of regularities might emerge and how they behave. However, the model can only elucidate these questions through a series of appropriately designed experiments. The experiments must be well thought out, with all the scientific rigor possible, and take into account several aspects of *in silico* experiment design, as well as several practical aspects of experiment setup.

3.8.1 Experiment Design Aspects

The experimental design starts with selecting the right type for the hypothesis, choosing a suitable time frame and defining scenarios and the scenario space.

3.8.1.1 Hypothesis Type

Given the generative nature of agent-based models, there are two main types of hypothesis we can form:

[14]This again shows the importance of using clear variable names ....

1. Under the specified conditions, a macroscopic regularity of interest emerges from the designed agent-based model.
2. A range of clearly identifiable emergent behaviours and regularities can be established from this agent based model of a system.

The first hypothesis relates to an attempt to model a real-world observed regularity. These models take a phenomenon that we see, but cannot explain, and attempt to provide an explanation of how we think it might work. This type of hypothesis is falsified if the model does not do produce the expected regularity, and is not falsified if the desired regularity emerges. Further, this type of hypothesis can ask questions about the conditions needed to produce the regularity, perhaps in order to explain why the regularity is observed in some real-world contexts but not in others that seem similar.

This hypothesis type will most likely have a fairly clear set of parameter values because they will be based on matching the real-world parameters in the context of interest. After completing the verification step, we should also have a good idea of which metrics to collect. Combining these two features of type 1 hypotheses means that experiments will be focused on examining these metrics over time to clarify if the regularity emerges and when. The experiments may also examine whether the regularity is stable or not, whether there are any concurrent behaviours of interest, or what might be the future behaviour of the whole system under investigation.

The second type of hypothesis does not attempt to recreate the real world so much as explore in what possible worlds we might find the regularity of interest, what else happens concurrently, and whether there are any parameters that delay, alter or disrupt the regularity. Type 2 hypotheses are oriented toward exploration, and do not necessarily have a "correct" or "incorrect" answer, but would be falsified if the regularity does not behave as desired at all, only behaves as desired under unreasonable conditions, or behaves chaotically, emerging or not emerging with no relation to the parameters under investigation. With no right answer, this type of hypothesis requires the modeller to identify whether behaviour is emergent and relevant when there is no clear guide on where to look for such behaviour. This means that there might be several subhypotheses (e.g. that a given parameter setting will completely prevent the desired emergent behaviour, or that the combination of two parameter settings will speed up the appearance of the regularity of interest), and each of these subhypotheses would be falsified independently. Unfortunately, there is no standard answer as to how to create such a hypothesis or how to test it, as it depends on the system being modelled and the specific questions the modellers and stakeholders have.

This second type of hypothesis will not have well-defined conditions to replicate, and instead is usually characterised by parameter sweep experiments with a (large) number of experimental model runs at varying combinations of parameters, whose value might change over time as well. We will still have an idea about the metrics of interest following the verification step, but since we do not really know what we are looking for, we need to consider that what appears to be too trivial to measure might actually reveal interesting model behaviour, either at each tick or at some predefined end point.

3.8.1.2 Time

A computational experiment with an agent-based model must be clear on how long is long enough for the experiment to be "done". We want to observe the behaviour of emergent patterns, but we cannot always know in advance how many time iterations are required for the formation of a pattern. Furthermore, any emergent pattern that arises might be unstable or oscillatory, may be sensitive to initial conditions so that it may arise easily and quickly, or slowly and only after much difficulty, or not at all, even under the same experimental parameters.

Type 1 hypothesis models may have an advantage in that replicating some real-world conditions will often provide some real-world timeframe that should also be replicated. If, for example, we are simulating the evolution of investment portfolios in a power generation sector as observed over the last 50 years, and the simulation progresses in ticks that represent quarters, we obviously need to stop the simulation at or around 200 ticks. In these models, the hypothesis changes from "Does the regularity emerge in this model?" to "Does the regularity emerge in the model's time limit?"

But not all simulations have such well-defined time frames for the emergence of a pattern. Many models investigate behaviour that appears continuous, punctuated or chaotic, so that each simulation tick has no meaningful time equivalent. Other models will not have a real world timeframe to replicate or are seeking to explore how much time would be required for the regularity to emerge under some conditions of interest. In these cases, the decision on when the simulation experiment should be stopped must be determined practically and empirically. A small LHS sample across the parameter space (see Sect. 3.8.2) left running as long as practically possible could yield thousands or even millions of time steps and that should be sufficient to able to explore whether and when expected emergent patterns appear, and whether observed stable attractors really are stable. Using this preliminary analysis as a rough guide of time requirements, we can make a more informed choice about the time limits to set on the remaining experimental setup.

3.8.1.3 Scenarios and Scenario Space

Finally, before exploring different types of experiments, the difference between a scenario and a scenario space must be clarified. Type 1 hypotheses are linked to scenarios, which, in the context of an agent-based model, can be understood as the set of real-world situations modelled and a narrative of what can or should happen in them. These scenarios might also look like a dynamic profile for some parameter set over time, relating the behaviour of the regularity to some variable that we see at different levels in different real-world conditions. Scenarios have a long tradition in modelling and are widely used, but must be approached cautiously because they can be limiting and may reflect wishful thinking about the system. Poor scenario selection can unrealistically simplify the variation in real-world conditions, resulting in predetermined results that fail to really explain the nature of the regularity. Furthermore, defining only a handful of scenarios (e.g. low, medium, high) might obscure some important relationships than a less limited set of scenarios would reveal.

As an alternative to selecting only a few scenarios, some type 1 hypotheses and all type 2 hypotheses should use a scenario space, where each parameter that can vary in the model represents a dimension, and where each point in that multidimensional space is an experiment. By examining the total scenario space, likely and unlikely combinations of parameters are tested, and may reveal surprising model behaviour or crucial thresholds after which behaviour changes. Time must also be considered as a dimension along which other parameters may change as well. Defining a scenario space reduces the chance of being locked into limited scenario thinking, although it comes at the hefty price of a vastly increased number of experiments that need to be performed. Depending on the complexity of the problem and the aim of the simulation, this may be a worthwhile trade-off.

3.8.2 Experiment Setup

After wrestling with difficult questions on the design of the ideal experiment, we now turn to addressing the practical limitations of experimental setup, including how many unique experiments to run, how many times each ought to be run, and how to deal with randomness.

3.8.2.1 Full Factorial

Full factorial experimental design is the most straightforward manner of performing a parameter sweep, where each experiment will be a point in a multidimensional parameter space. While full factorial design is rarely used for problems with more than five parameters (Wu et al. 2000), it is conceptually the simplest approach, and will be used as the foundation for the discussion on computation setups below.

Full factorial experiments define the smallest and largest possible values for each parameter, as well as the step size by which we increase from the starting value as we test all possible combinations for all possible parameter values in all possible dimensions. While a full factorial setup is logical, thorough and easy to setup, the number of experiments can quickly become too large to handle. For example, let's assume we model an industrial network with variable parameters for the prices of the goods the cluster produces (see Nikolic 2009). Testing each price at pseudo-logarithmic intervals of $0.1, 0.2, \ldots, 0.8, 0.9, 1.0, 2.0, 3.0, \ldots, 8.0, 9.0$ and 10.0 for 14 different good we produce 20^{14} combinations of goods-prices. If each simulation run takes 1 second, then simulating all these possibilities would take 20^{14} seconds (5.2×10^{10} years), more than 10 times longer than the estimated age of planet Earth (4.6×10^{9} years). For this model, a full factorial experiment is clearly infeasible.

The number of experiments needed for a full factorial experiment can be decreased by reducing the granularity of our simulation or the range over which the parameters can vary. This may be enough to bring the total number of experiments below a practical limit, although it may reduce the resolution of the experiment as well. In the case of the industrial network model above, reducing the price intervals to factors of 0.2, results in 10^{14}, which is still too many. Full factorial experimental setup might work for a small number of dimensions or values in each dimension (e.g. up to 3 or 4), but a larger scenario space demands a different approach. Furthermore, traditionally, factorial design of experiments is used to estimate the model behind a set of measurements, and not explore a model across a wide range of settings. Because of this the statistical instruments normally used around factorial designs is not suitable for our purpose. Alternatively, one could use fractional factorial design technique (Wu et al. 2000), but its statistical structure breaks when large number of parameters (above 30 for example) are explored.

3.8.2.2 Random Parameter

If the full factorial parameter sweep is too large, then we need to find a feasible maximum number of experiments that we can afford to run. Using this maximum as a guideline, generate a set of randomly distributed parameter values in each dimension by drawing x samples from a uniform distribution between a minimum and maximum value for the parameter. While seemingly rational, the actual distribution of parameter points achieved this way is not as uniformly distributed as one would have hoped, so random parameters should be limited to preliminary experiments or to experiments with few but wide ranging parameters. Figure 3.4 shows a histogram with the frequencies of samples falling within a range of values, and demonstrates the difference between samples generated by a random sampling and those generated by a technique called “Latin Hypercube Sampling”.

3.8.2.3 Latin Hypercube Sampling

Latin Hypercube Sampling (LHS) is a statistical technique that guarantees uniform sampling with the desired granularity of the scenario space given a Y dimensional parameter space and with a limit of X experiments. Unlike random parameters, which are chosen by considering only one parameter at a time, LHS considers the entire set of parameters and finds where in the parameter space we should perform the predetermined number of experiments to get the most representative subset of the space. Software tools including Matlab and R have built in functions for generating LHS samples. Although easier than a full factorial experiment, large sets of LSH samples (for example 10000 samples from a 30-dimensional distribution) are not computationally trivial and may require large amounts of computer memory and CPU time. The interested reader is advised to examine the work of Tang (1993) and Ye (1998).

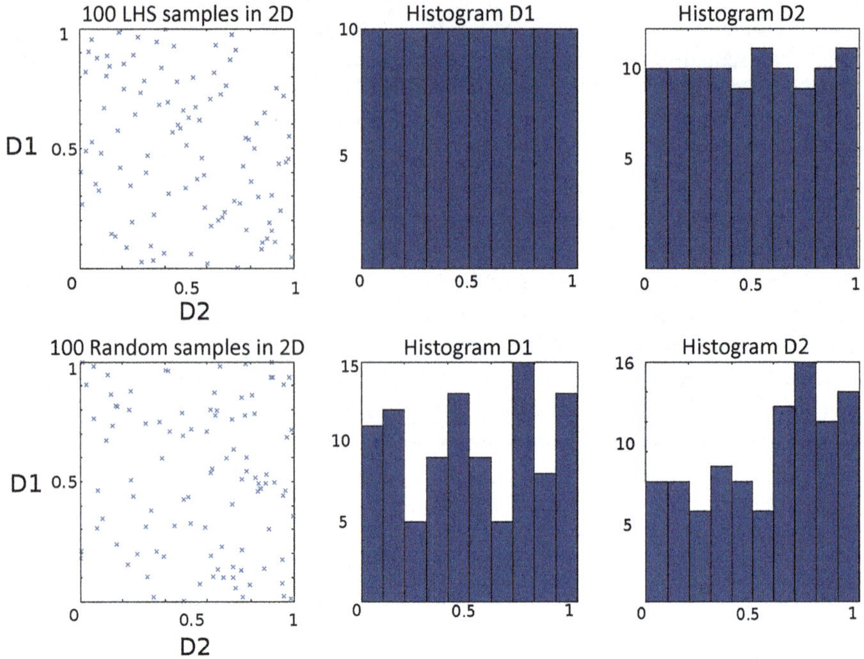

Fig. 3.4 Latin Hypercube Sampling (*top row*) vs Random Sampling (*bottom row*)

3.8.2.4 Monte Carlo

Monte Carlo experiments are a family of related approaches to finding an appropriate subset of the scenario space to test. It generally consist of three stages. First, we define the parameters that we want to explore and their expected values. Second, we generate random parameter values around the expected values using certain probability distributions and the degree of uncertainty on the parameter values in order to choose the width of the distribution and its shape. A wide range of possible values and a uniform distribution will essentially recreate a random parameter sweep. Third, we need to perform model experiments at the generated points. Extensive discussion on Monte Carlo methods can be found in Robert (2004).

3.8.2.5 Repetitions

Given that agent-based models are chaotic due to their iterative nature, a golden rule of agent-based experiments is to "never trust the outcome of a single run of an agent-based model" because every single run could be unrepresentative outlier. Reliable statements about model behaviour demand multiple runs and descriptive statistics over the total set of runs. How many runs makes a statistically significant sample? And is that same sample size sufficient for every point in the scenario space or will a larger sample be required at certain parameter values? There are no right answers, but a knee jerk reaction would be to just do many runs, perhaps 100, for each experiment. The need for a good sample size must be balanced by the practical need to finish the experiment within a reasonable time. Note that if a set of experiments take a full day to run, then 100 repetitions takes three whole months, which may or may not be too long for the project.

In order to find a good balance between reliably large samples and reasonably long experiments, we propose the following strategy. First, perform 100 repetitions of a small number of experiments distributed across the parameter space using LHS. Perform descriptive statistics on these points to explore the variability of the outcomes and identify the most variable metrics and parameter values. A discussion on estimating the number of samples needed for a given confidence level may be found in Bluman (2004). When you have found the sample size needed for the most variable metrics and most variable regions of the scenario space, err on the side of caution, and performing that number of repetitions for every experiment.

3.8.2.6 Random Seed

Even in cases where agent behaviour is fully deterministic (i.e. it contains no random elements) a simulation may still contain stochastic elements, as the agent iteration is often randomised in order to prevent first-mover artifacts. This means that each model is partly governed by a pseduo-random number generator, introducing two issues to be aware of: the possibility of pseudo-random algorithm cycling and the ability of exact run replication.

Depending on the exact implementation of the pseudo-random number generator, the pseudo-random number sequence could theoretically repeat itself. Most modern pseudo-random generators use the Mersenne Twister algorithm (Matsumoto and Nishimura 1998), which will only repeat itself every 4.3×10^{106001} number generated, but older algorithms may cycle more often and influence the outcomes. In cases with particularly large number of experiments, or those that take many timesteps to compete, it may be relevant to check the pseudo-random algorithm used.

Another important aspect of pseudo-random number generators is the need to produce the exact same sequence of "random" number in order to replicate results by fixing the *random seed*. Each pseudo-random algorithm requires a seed number from which it starts its calculation, usually the current time expressed in milliseconds since a certain date in the past, and this seed can be stored in order to replay the algorithm later and get the exact same sequence out. Saving the random seed is especially useful for when working with a type 2 hypothesis because, although the original experiments may not be looking for anything in particular, if something interesting appears in the output then the modeller may want to re-run a specific experiment to record additional metrics.

If such post analysis is likely, check if the model code inadvertently calls multiple pseudo-random generators during its execution. Java, for example has a built in random generator that can be used, while the modelling environment Repast might use another from a separate library, necessitating that both random seeds be stored. Furthermore, Java (and other programming languages as well) has several data structures, such as the *HashMap*, that do not have a predictable iteration order, bringing additional randomness to the model that must be accounted for to have perfectly repeatable simulation run. As a simple test, repeat a simulation 100 times with a fixed random seed and compare the outcomes to reveal whether the model is fully repeatable or not.

Remember that any change to the order at which the random number generator is called, for example if the code governing agent behaviour is altered, will not give the same outcomes as before these changes in the model because the order in which the random numbers are drawn may be different.

3.8.3 Experiment Execution

Compared to traditional experiments in a lab, running an experiment with an agent-based model may be underwhelming. After designing the experiment and dealing with the practical limitations, the execution is just running a computer program and recording the output. But performing many thousands of experiments might create very large amounts of data and high dimensional parameter spaces (20+ parameters) can equate to several hundreds of thousands of runs, potentially generating hundreds of gigabytes of data. This would require nontrivial[15] amounts of computational and storage resources and presents a further set of limitations on the experimental execution.

3.8.3.1 Running on a Single Computer

If the only computer available is the modeller's desktop or laptop machine, then many days of continuous computation may be required. This may not be a problem for a desktop, but laptops are usually equipped with low power CPUs and inadequate cooling for running many hours under full CPU load, which might cause overheating, automatic shut-downs, or, even permanent damage to the hardware.

If multiple computers are available, consider splitting the experiment to smaller parts, and running these parts on multiple machines through distributed systems such as BOINC.[16] BOINC, made famous through SETI@home project for the analysis of incredibly large volumes of radio signals in the search for extra terrestrial intelligence, can take advantage of underutilised office computers. Even though setting up a BOINC system requires specific ICT skills and cooperation form the system administrator, it is worth considering if running the experiments on a single computer does not suffice.

[15] At the time of writing.

[16] See http://boinc.berkeley.edu/.

3.8.3.2 Scaling, Running on Clusters

For running large scale computational experiments the ideal situation is to employ a computer cluster, as even modest clusters can have a computational capacity several orders of magnitude larger than desktop PCs. If large parameter sweeps and many repetitions mean that the total run time on a PC is unreasonably large, then preparing an experiment on a cluster may be less time consuming, even if it is more complicated to set it up initially. In our experience, a rule of thumb is that it is worth looking at using a cluster if the PC computation time is longer than a few days on a regular desktop or a single server.

Simulations on a high performance computing cluster can be performed in two different ways:

Many single runs: In this setup, each cluster CPU get a single model run assigned to it, and when the model run is finished, a new one gets assigned. This is easy to set up, however, a single run must be executable on a single CPU within reasonable time and require no more memory than available per CPU.

One model on clusters: Here, a single model run is distributed across the cluster, with each CPU or node containing a single (or a small set) of agents. This is far more difficult to set up, as the actual model code has to deal with agent communication issues, network latency, shared information etc. This is an option if agents do not communicate a lot, but have CPU intensive reasoning that uses a lot of memory. An example of such a case is a model where multiple geographical regions can be simulated on different physical machines.

3.8.3.3 Collecting and Storing Data

Models commonly use separate text files to store the data resulting from individual simulations. This way, if any of the many simulation runs fails, the data from successfully completed simulation runs will not be affected, although the large number of files this process creates may be unwieldy to deal with.[17] Parallel simulations that write to a single output file may experience *concurrency* issues if multiple simulations try to write to the same file at the same time. Furthermore, it is crucial that each output row in the file must contains information about the experimental conditions, the simulation run and the tick so that the output can be unequivically matched to the specific simulation run from which it was generated.

Simulations often collect several different types of data, such as the state of individual agents, network connections between agents or the overall characteristics of the simulated system. A back-of-the-envelope estimate will show that multiplying the data collected from a single run by the number of experiments in the scenario space, by the number of runs needed to get a significant sample may quickly go

[17] Please note that older file systems, such as FAT16 or FAT32, often used on USB disks, can only handle around 65000 files.

beyond the storage or data analysis capabilities of software on a personal computer, requiring other solutions, such as a database. Using a database for storing simulation data has two main benefits. First, powerful *query* languages such as SQL allow efficient navigation through the data in databases. In practice, this may significantly reduce the amount of analysis code written by concisely stating searches without having to code the algorithms as would be necessary in a programming language. Second, databases do not load all of the data into the working memory at once, instead filtering data to get elements of interest that fit into memory and they are designed to deal with large quantities of structured data.

3.8.3.4 Checking Data Consistency

When we perform a large number of computational experiments there is always the possibility that something went wrong with at least a few runs. If we have distributed our simulation across many machines, some might have overheated and crashed, or if we performed the experiments on a cluster, it might experience a network overload. Therefore it is always wise to check whether all experiments have finished cleanly and whether we have stored all data correctly. Depending on how we stored our data files (separate data file, one large file, a database etc.) we might need to do different things. In general, we can predict exactly how many data rows we might expect (i.e. $NumberOfRows = NumberOfExperiments * NumberOfRepeatedRuns * NumberOfTimeStepsRecorded$) and verify that we have all the data. If we come up short, we need to analyse the recorded parameter data and compare them with our experimental design. It is advisable to automate such testing with scripts, and generate reports for missed experiments, as the rerun may also give us problems.

3.8.4 Step 7 Example

Hypothesis Example We are interested in using the model to study technology diffusion in the Westland. We will focus on the technologies in use by the companies in our experimental setup by measuring the relative number of companies that use a certain technology. We want to ask whether we get an emergent pattern of technology diffusion, and what parameters affect the pattern of technology diffusion?

We could vary the characteristics of the technologies (e.g. price, costs, production, lifespan) or the variables that influence the opinion formation (e.g. stubbornness, degreeOfNeighbours), or both. For the first experiment, we fixed all the parameters in order to examine the diffusion process. The 8th technology in technology category 1 has the best effect on production, even though it has a high purchase cost, so we hypothesise that over 250 ticks more companies will purchase this technology than the others in the same technology category (type 1 hypothesis).

The second experiment will explore how some parameters affect the diffusion process (type 2 hypothesis). We will use LHS (see Sect. 3.8.2.3) to select a subset of experiments over wide ranging Stubbornness and OpinionChangeRate parameters to investigate how these variables influence the diffusion process alone and in combination.

Experiment Design—all settings fixed to focus on the diffusion

- General setup
  - Ticks [250]: Each tick = 1 year. 250 years is long enough to explore possible convergence on the "optimal" technology
  - Repetitions [100]: Enough to get a good sample of results, not unfeasible for the hardware setup
- Parameter settings
  - NumberOfCompanies [50]
  - NoiseChance [0]: Not tested.
  - Budget [0.3]: Set maximum budget allowance per technology purchase at 30 %
  - LocalProfits? [true]: Satisfaction is derived from comparing profits to that of neighbours
  - UnlimitedFunds? [false]: Companies can go bankrupt and be reinitialised
  - Visualisation? [false]: Visualisation turned off to reduce time and computing resources.
  - DegreeOfNeighbours [2]: With 50 agents in total, a degreeofneighbours of 2 means each agent knows about 1/4 of the others.
  - Stubbornness [0.95]: Agents value 1st hand knowledge strongly over information from others
  - OpinionChangeRate [0.2]: Equates to quite a big memory, which buffers against sudden or big changes in opinion
- Output at each tick
  - The tick
  - Random seed used (to be able to reproduce the results)
  - Surface area of companies
  - Type of companies
  - Broke counter of companies (number of companies that have gone broke and been reinitialised)
  - Credit balance of companies
  - Satisfaction of companies
  - Technologies of companies (variable of most interest).

Experiment design—Parameter sweep experiment

- Setup
  - Ticks [500]: More time to allow for disruptions to any possible convergence
  - Repetitions [50]: Fewer repetitions, although still a reasonable sample, in order to reduce analysis burden
- Parameter settings
  - NumberOfCompanies [50].
  - NoiseChance [0]: Not tested.
  - Budget [0.3]: Set maximum budget allowance per technology purchase

  - LocalProfits? [true]: Satisfaction is derived from comparing profits to that of neighbours
  - UnlimitedFunds? [false]: Companies can go bankrupt and be reinitialised
  - Visualisation? [false]: Visualisation turned off to reduce time and computing resources
- Parameter sweep
  - Stubbornness [0.5–0.99]: Wide ranging sweep using 50 LHS samples
  - OpinionChangeRate [0.01–0.5]: Again using 50 LHS samples
- Output at each tick
  - Random seed: to be able to reproduce the results
  - Type of companies
  - Technologies of companies (variable of most interest).

3.9 Step 8: Data Analysis

It may seem that once we have experimental results, we are on the home stretch, with "just" the analysis to be done. However, the data analysis is arguably one of the most difficult parts of the modelling cycle. We are embarking on a adventure in "experimental data land", where no one has gone before, where there are no maps and where we do not know what manner of weird and wonderful creatures we will meet. Data analysis requires a truly open mind and an exploratory spirit.

The experimental design, as described in the previous step, included making decisions on which model metrics to record. Now, after running all the experiments we need to sort through all those recorded metrics in order to answer the question formulated in Step 1 (Sect. 3.2). We must keep the story we want to tell or the problem we want to solve in mind, as we proceed through the following data analysis phases:

1. Data exploration
2. Pattern visualisation and identification
3. Pattern interpretation and explanation
4. Experiment iteration.

The process works, in general, by first exploring the data and identifying interesting or relevant patterns, visualising these patterns and interpreting them in order to understand the generative processes that created them. Finally, this interpretation will—more often then not—lead to new questions requiring additional experimentation.

3.9.1 Data Exploration

The data analysis begins with data exploration. Assuming that we have verified the data consistency, we can start exploring the outcomes. We essentially ask ourselves "what do we have here?" We need to identify the relevant emergent patterns for answering our original research question. Finding (the absence of) that pattern will answer the research question. However, how do we know where to look or how to recognise a pattern when we find one?

3.9.1.1 Hypothesis Driven Analysis

When looking at the results, modellers may wish the compare a specific run to another specific run, or they may want to compare the statistics derived over multiple simulations run under the same parameter setting compared to those derived from another parameter setting.

Depending on the experiment, modellers may wish to compare the entire simulation run, or only a select number of states, such as the end state, but either option presents the same basic comparison: are the differences in results from one case to the other statistically significant? Experiments based around a type 1 hypothesis (see Sect. 3.8) are likely to find that the multiple runs of a single experiment all have a similar set of results, such as that the emergent regularity of interest does or does not appear. A look at the entire run might reveal more detailed patterns, such as that the regularity tends to appear around the same time in each run, or that it will only appear after, in conjunction with, or in the absence of another pattern of interest, or that after first appearing, the pattern takes some similar amount of time to fully stabilise. Finding these types of patterns requires thorough use of descriptive statistics to know whether a given set of results can be judged as the same or similar, or if in fact they are significantly different despite superficial similarities.

On the other hand, falsifying the more exploratory type 2 hypotheses will require looking to see under which experimental conditions the results are different, and if these differences hold consistent relationships to the conditions. Finding these patterns is more difficult, because the modeller must first describe the results for each unique experiment as if it were a type 1 hypothesis, then compare the results for the different experiments to see if they are different. Modellers will therefore need to compare the variance between conditions after describing the results of each condition.

3.9.1.2 Location of Patterns

The parameter space and the data output of an experiment are usually too large to examine all possible relationships between them. We need to narrow down our search to places that are likely to hold meaningful patterns. The metrics we recorded and are now studying usually fall within three groups, namely agent states or behaviours, agent interactions and collective agent states and behaviours.

Firstly, patterns may be detected in the states or behaviours of single agents. It might reveal that, for example, a certain agent is unable to make a profit and goes broke during the course of the run, or that an agent producing a certain kind of good improves production efficiency only if supply is greater than demand.

Secondly, patterns may also be detected in the interactions between agents. For example, when agents move from a regularly connected network to a small world network, they make and break links with other agents, and the pattern of changes may be interesting. Similarly, agents may begin with highly disorganised bargaining strategies but quickly refine them, so examining how many fewer rounds of negotiation are needed by the end of the run can reveal the effectiveness of the learning strategies.

Lastly, patterns may appear in collective or aggregate agent states. Interesting aggregate states might include the total CO_2emissions from the technologies operated by the agents, or the average tax payed by them, and the change in these states over time may demonstrate how decision making works in the model or how effective a policy is.

3.9.1.3 Data Analysis

At this point, we must revisit our golden rule of modelling, defined in Sect. 3.8, about never trusting a single run of an agent-based model. Particular care must be taken to examine repeated runs at the same settings, as we might discover that the run is at a metastable region of the parameter space, and different runs end up in different attractors. We first need to isolate the repeated runs from the entire set, and for this subset establish a number of metrics to look for sets of tendencies in the multiple runs of each unique experiment.

We may want to calculate the average and median values for each metric, either at the end state or at each time step for time series experiments. Furthermore, variance and the "kurtosis" (i.e. the so-called peakedness of a distribution) should be calculated to detect bifurcations within the repeated experiment and to get an idea of the variation of the data generated from the experiments. Outliers need to be identified, although not removed, as they might indicate instabilities in the emergent pattern. The values of variance, kurtosis and other measures plotted against time might reveal useful insights. Once we have calculated a coherent description of each unique experiment by combining the repeated runs, we need to compare different unique experiments with each other using normal statistical methods, such as ANOVA ("ANalysis Of VAriance") tables.

At this stage we are looking in the data for something that may not exist, or may exist but could take an unexpected shape or be in an unexpected place. When exploring emergent patterns, we are in constant danger of overlooking important clues by focusing the search on something else. Blinkered by expectations, one might miss important conclusions about the model.

3.9.1.4 Analysis Tools

Recognising patterns is something that humans do naturally. Our brains (i.e. our neural networks) are naturally wired to perceive structure and patterns, even when there are none. Consider laying in the grass looking at the sky, constantly spotting new but familiar shapes in clouds above us. We are wired to say "Hey, interesting!" or "Ha, that looks like something I've seen before!". Computers, on the other hand, do not know what is interesting to the user. When we do not know exactly what we are looking for, we must embark on an iterative process of examining the odd or interesting, in order to discover something more interesting that is causing it. Software tools can help us manage the large amounts of experimental output so that the user can explore the data in a search for patterns (see also Sect. 3.9.2).

A wide array of software tools is available to aid data analysis, and the choice will, among other things, depend on how data was stored (e.g. in databases or in "flat" files). We will discuss four tools here, namely Excel, R, SPSS and Matlab/Octave:

- Excel, while widely available and a familiar tool to most computer users, has a main drawbacks in that its syntax is difficult to debug and that older versions have limited capabilities to handle large amounts of data. It may be sufficient for smaller projects, but for more complex tasks it is worth considering alternatives that are better suited.
- R is a de-facto standard in the statistics community. It is open source and free[18] and it has a lively community of developers and active users, making it easy to learn, even for the self taught, and relatively easy to get support online, even for seemingly unusual or obscure problems. It also makes the results more transparent for those wanting to repeat research. R allows you to define your own functions and there are many packages available to extend it, for example for advanced visualisation. Agent software packages such as Repast Symphony provide a direct link with R from the modelling user interface, so that the output data of the model is instantly available in the right format and ready to be analysed.
- SPSS is a closed source and commercial package that is popular for statistics in social sciences. While powerful, SPSS is not always intuitive to use. If SPSS is available and the user is already familiar with the software, it can be an excellent tool to analyse the model outcomes.
- Matlab is a general purpose matrix computational engine. It is powerful, extensible and also has a large user base. However it is a commercial product with a significant price tag and it may be an overkill for basic data exploration. Again, if the software is available and the modeller is skilled in its use, then Matlab will provide a valuable tool for data analysis. Octave is the free open source counterpart to Matlab, which in many cases can act as a replacement. While not as feature rich as Matlab, for many applications it will more than suffice.

[18] See http://www.r-project.org/.

3.9.2 Pattern Visualisation and Identification

After data exploration, the next phase in this step is pattern visualisation and identification. We will address recognition of patterns and visualisation as well as several pitfalls.

3.9.2.1 Recognising Emergent Patterns

Aided by the software analysis tools, the user must recognise emergent patterns in the results of the experiments. The types of patterns we expect to find in data (either over time, or over some parameter value) are:

Dynamic Behaviour There are many types of dynamic behavioural patterns, including cyclical, exponential and power laws. Identifying dynamic behaviour can be difficult if the experimental setup is insufficient to reveal the full pattern. For example, the frequency of the cycles or the ramp-up of an exponential growth may be too long to be visible in a short simulation run or if the parameter sweep is too limited. The frequency of cyclical behaviour can also vary over time or the dynamics may be chaotic and difficult to recognise.

Attractor Changes We might observe sudden jumps or collapses as the model goes through an attractor change. Again, the issue of time is important, as the transition will seem more severe in a longer time run or in parameter sweeps with different granularity. For a discussion on attractors, please refer to Chap. 2.

Metastable Behaviour When repeated runs are examined we might observe that runs are clustered in, for example, a low and high set.

Lack of a Pattern We might observe randomly noisy behaviour where the data does not seem to follow any pattern. Alternatively, we might not see any change at all across time within the parameter space.

The main way we recognise these patterns is by graphically displaying them because it is easy to overlook seemingly insignificant details in the raw data.

3.9.2.2 Visualisation

There is no single approach to visualising patterns in experimental results. Instead we need to explore multiple perspectives on the same data, examining each critically to determine which yields the most insight into the problem. Data visualisation is really more "art than science", but even a non-artist will improve data visualisation techniques through a solid understanding of the system modelled and the questions motivating the research. Figure 3.5 provides a handy starting point to deciding how to present the data. We would like to add network representation and animated visualisations to the visualisation options presented in Fig. 3.5.

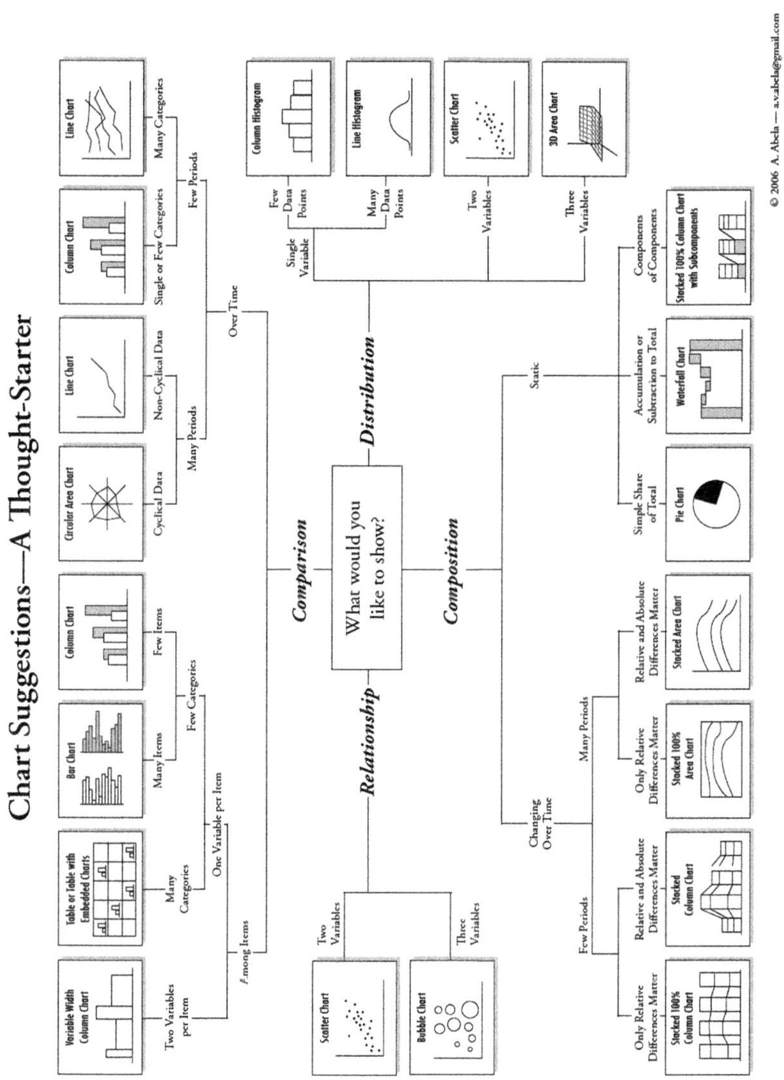

Fig. 3.5 Different types of visualisations (Abela 2008, 2011), used with permission

Network Representation The options under the "Relationship" arm of Fig. 3.5 are about relationships between data. Relationships between agents belong here as well, so we would like to include graph topologies in various layout options and other network metrics. Options for graph layouts include tree maps, circular layouts and force directed graphs, among others. We might also want to examine the agent social data (such as the structure of money flows between agents) or the physical network (e.g. the flows of mass between the physical nodes). When dealing with network data, we can use additional network metrics to understand them, such as degree distribution, average shortest path lengths or clustering coefficients. See the work of Newman (2003) for extensive discussion on network metrics.

Animation While requiring extra effort, animated images can be insightful and serve as a great communication tool. All of the visualisation options discussed above can be animated by adding an extra dimension, such as time, to create video sequences of metric changes. The data analysis software mentioned above can create a series of images across the extra parameter which can be merged with video editing tools. Open source tools such as ffmpeg[19] or mencoder[20] are free and produce professional results. Alternatively, one can use graphing libraries, such as Prefuse[21] or Processing[22] to create interactive data visualisations of any data set. This approach requires more advanced programming skills, but the results can be truly stunning.

3.9.2.3 Visualisation Caveats

Graphs are more than just pictures. They are highly condensed packets of information and the way that information is condensed can greatly affect perception and interpretation. This can be advantageous, if the concentrated information is clear, coherent and obvious, but the process of putting the information into a graph can also make it obscure, inconsistent and misleading.

The choice of metric and statistics shown greatly affects how the message is perceived. Absolute values can tell a different story than relative values, just as means presented with or without a standard deviation can lead listeners to widely divergent conclusions. In addition to the choice of information presented, the choice of presentation can also affect its reception because the graph type can reveal, hide, or distort the information. For example, something twice as big on a *spider chart* appears as a quadratically larger surface, while the bin size in a *histogram* can affect the pattern heavily. Using logarithmic axes will make the depicted relationships appear more linear, the orientation of axes can suggest unsupported trends, and axis scaling can greatly influence the perceived magnitude of the pattern. Even colour choice can

[19] http://www.ffmpeg.org.

[20] http://www.mplayerhq.hu.

[21] http://prefuse.org/.

[22] http://processing.org/.

subconsciously bias the viewer in interpreting figures, from culturally based effects for "good" and "bad" colours to using colours which are difficult to distinguish from one another.[23] Finally, a complex graph may be so difficult to interpret that it requires a lot of explanation, defeating the purpose of graph, and alienating the intended audience. All of these issues, and potentially many more, could significantly influence the way information is *perceived* by the recipient, and the conclusions drawn.

3.9.3 Pattern Interpretation and Explanation

Once we have identified and visualised the patterns that we expect will help us answer our research questions, we need to interpret them and provide an explanation for their emergence, confirming and rejecting our hypothesis. To interpret a pattern means to make sense of it, define it, or translate it to another language. Every bump on the graph needs to be understood and explained. Every observed change must have a justification. Every little part of the pattern must be depicted not only visually, but also logically.

In practice, this means creating a story that tells how the interactions and conditions have lead to the observed patterns. Put another way, the story must tell what is going on and how we know what is going on. It should be coherent enough to know which interactions, or which combinations of interactions, are crucial and which are helpful but not necessary, merely incidental, or even detrimental. Likewise, the story should identify which conditions will inevitably lead to these outcomes, which might increase or decrease the likelihood of such an outcome, and which will inhibit them.

The explanation should relate directly to the experiments performed and the hypothesis that drove the experimentation. The interpretation must make it clear how the experimental conditions have changed the patterns, as well as whether the changes are direct results of the varied parameters or instead an indirect result of how the parameter has changed something else.

3.9.4 Experiment Iteration

After interpreting and explaining the observed patterns, it may be necessary to go back to the experimentation step (Sect. 3.8) to redesign the experiments and run the model again, taking into account everything learned so far. There may be interesting patterns that warrant a closer look, or possibly emergent behaviour which requires further experiments for a less ambiguous interpretation. To clarify or further explore identified patterns, one can consider:

[23]This is especially true for the approximately 10 % of the male population who is colour blind.

- Broadening the parameter sweep;
- Increasing sweep resolution; and
- Introducing new metrics.

Broadening the parameter sweep can be considered as "zooming out" and looking at the system from a wider perspective, and can include increasing the ranges within which the parameters are varied or varying additional parameters that were held constant in the original set of experiment. Broadening the parameter sweep in this way might entail reducing the density of the samples to make the simulation runs manageable (see Sect. 3.8.2). On the other hand, increasing the resolution allows one to "zoom in" to one specific area, by taking more samples within a smaller range or with fewer parameters. Running additional experiments in a smaller area can help to identify the cause of the patterns you observed, to find out if the pattern actually exists as you think it does, or to identify the "sweet spot" at which behaviour changes. Finally, after running a set of experiments, it may become clear that new metrics should be introduced to the simulation in order to measure new behaviours, measure the original behaviours more accurately or to measure new features of the original behaviours that may better reveal the underlying explanatory variable.

It must be stressed that after re-defining the experiments, the choices for visualisation of the data should be revisited. The previous visualisation choices may have worked quite well, but new experiments mean new information and the same steps will have to be taken to ensure the approach is suitable.

3.9.5 Step 8 Example

Hypothesis Driven Analysis The hypothesis in this data analysis example is of type 1, i.e. an attempt to replicate an observed real-world phenomenon in a simulated environment. The hypothesis itself is that diversity and adaption, as seen in biologically evolving species, is also present in non-biological entities, in this case greenhouses. Finding that non-biological entities diversify and adapt would strongly support the idea that non-biological entities are evolving by some of the same mechanisms as biological entities.

All the greenhouses are initialised with an identical 0 technology for all categories, with no cost and no effect on production, meant to represent an absence of technology. All other technology options have varying costs and varying effects on production, with only one optimal technology per category and per crop type, all others costing more or producing fewer crops. This hypothesis would be *falsified* if the simulated greenhouses showed no evidence of diversifying or adapting their technology selections, showed no preference for those technologies that performed better, or showed the same distribution of technologies regardless of crop type. Therefore, the expected pattern would be that the greenhouse company agents will significantly favour those technologies that perform best for their crop type, even if these technologies are more expensive.

The distribution of technologies by crop type could be analysed at the end state, but the relative distribution of technologies over time could reveal interesting dynamics. The improvement in performance must weigh against the increased purchase and operating costs of the more advanced technologies, and these competing pressures might reveal conflicting behaviours, so we chose to record every tenth time step, from a total of 500 steps, to form a time series.

The output of the simulation included the technologies owned by each greenhouse grower at each recorded time step. Multiple runs were summed together, resulting in a total of 12,000 greenhouses, initially randomised to either flower (type 0) or vegetable (type 1) crops and with a 0, or absent, technology for all technology categories. The number of greenhouse company agents with a given technology in each category were added up to count the total number of examples of each technology in the simulation at each recorded time step. Separated by crop type, and plotted against time, these counts show how the proportion of technologies changes over the course of the simulation. Algorithm 3.7 shows and example of R code for plotting agent states over time, summing up 100 repetitions at the parameter stetting. Figure 3.6 presents the resulting graph.

Algorithm 3.7 Example R code for plotting the companies water technology ownership per timestep

```
test <- sqldf("select step, TOW, count(companyID) as
    companyCount, CompanyTypeOutput from adf where
    runnumber=1 and CompanyTypeOutput=1 group by TOW,
    step")
test$step = as.numeric(as.character(test$step))
test$CompanyTypeOutput = as.numeric(as.character(
    test$CompanyTypeOutput))
ggplot(data=test, aes(x=step, y=companyCount)) +
    geom_area(aes(fill=TOW)) + facet_wrap(~
    CompanyTypeOutput)
```

The graph in Fig. 3.6 shows us three things. First, that flower growers are initially far less profitable than vegetable growers (this was built into the model) so the number of companies growing flowers drops dramatically as these companies go bankrupt, to be replaced by another company which has a 50 % chance to be initialised as a vegetable grower instead. This disadvantage appears to decrease over time as the effect of technology on production brings the flower growers potential for profit more in line with that of vegetable growers. The difference in behaviour shows diversification at work, while the improved performance of the flower growers show adaption in the face of pressure to adapt or perish.

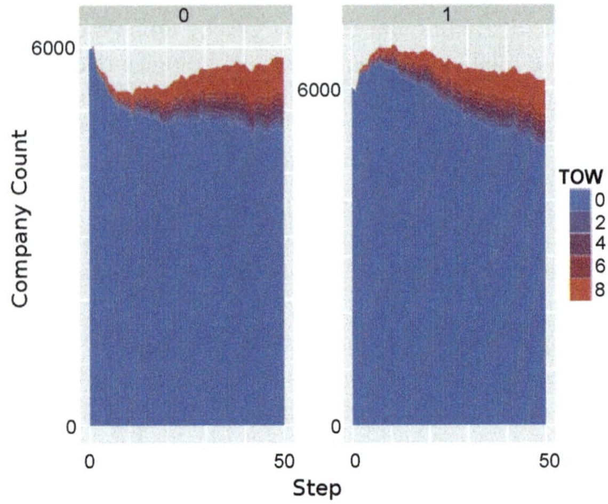

Fig. 3.6 The total count of companies with each type of technology in the greenhouse water technology category. Results divided by company type, 0 representing flower growers and 1 representing vegetable growers

Second, both company types quickly start increasing ownership of non-0 technologies as any technology choice at all is better for performance than the null option, showing adaption. Further, the distribution of technologies is not equal in both company types. Types 8 and 4 are good for both crop types, and are consequently popular for both. Less popular, and less equally distributed, are the technologies that are only good for one type. Technologies 7, 6, and 5 are, in decreasing order, good for the production of flowers but bad for vegetables, while the technologies 1, 2, and 3 are good for vegetables but bad for flowers. The uneven distribution of these one-crop-only technologies hints at the diversification of the companies by crop type. As these technologies are not evenly distributed, the hypothesis that the companies will diversify by crop type to find an adaptive advantage is supported, although it could be supported more clearly if the simulation were allowed to run further.

Third, although the distribution is not equal for the two company types, all technologies are owned by both types. The opinion sharing mechanism in the model does not allow the performance evaluations to include crop types, so agents have no way to know whether their neighbour's favourable opinion is due to an effect on production that they will share the benefits of. Agents therefore must experiment with all technology types themselves, and can be persuaded to buy a technology that has a poor effect on their production if their neighbours are overwhelmingly of the opposite crop type and are therefore making good use of the technology. This is not an error in the model, but could potentially make for a detrimental situation for one crop if the other were to become dominant (as does happen in real life).

3.10 Step 9: Model Validation

Whereas verification concerned the question "did we built the *thing right*" (i.e. whether the model implementation matches the design), validation addresses the question "did we build the *right thing*". Did we design and build what we needed to answer the questions of the problem owner? Are the outcomes convincing? This vital step ensures the experimental outcomes are worth something to the problem owner, but is best viewed as a continuous process, rather than a single check.

The traditional view on validation is concerned with whether the model is an accurate representation of the real-world system (Sargent 1998) and takes the shape of a series of comparisons between experimental results and real-world data. However, traditional validation methods are not always applicable to agent-based simulation (Louie and Carley 2008), especially those based on "what-if" type 2 hypotheses. Validation cannot simply compare computed behaviour to "real" system behaviour if there is no "real" system available for comparison or if the model is exploring possible future states. Instead, validation for this type of models focuses on whether the model is *useful* and *convincing* in its explanation of how a system possible operation or potential states of the system. The real outcome of the model is not the experimental results so much as an increased insight and knowledge and *that* outcome can be validated through several different methods, including:

- Historic replay;
- Face validation through expert consultation;
- Literature validation; and
- Model replication.

These methods are discussed in more detail below.

3.10.1 Historic Replay

For models that can be compared to a real-world situation, a historic replay consists of a scenario and computational experiments. The scenario is a description of the path from some past point to the state of the world today through the system's observed parameter space. The people, actions and situations from the scenario are parameterised and translated to the model, then experiments explore whether the emergent patterns from the observed scenario are present in the simulations and if the outcomes and end states of the model resemble the current state in the real system. If a pattern emerges in the computational experiment that corresponds to the trajectory described in the scenario of known situations, we can claim a degree of validity by the model. Of course, the word "corresponds" is problematic here. Do we expect an exact quantitative replication of some pattern, for example exact profits and losses of companies, or are we satisfied with a qualitative similarity of rising and dropping assets? The requirements depend on the specific case and problem owner, while other experts can help determine what is good enough (i.e. how close the outcomes should be).

There are some challenges and obstacles when preparing historic replays with your model. First of all, we may not know the state of agents in the past or the interactions that might have influenced those states or caused adaptations in agent behaviour. This data may be simply unknown, as nobody thought it would need to be collected, or it might be present but private and unavailable. Studying historic developments of socio-technical systems also shows how sensitive these systems can be to initial conditions or to unexpected situations. The initial conditions for a decision, including information available to the decision maker or option to choose from, would be different depending on whether or not the decision maker had just run into an old business partner in a pub and discussed the decision at hand. Likewise, external global incidents that influence everything but that could not have been predicted (e.g. the terrorist attacks on the 11th of September 2001) are especially difficult when looking at only the specific domain which we are modelling. Few supply chain models include international conflicts, but they actually have a large influence on the operations systems, for example through oil prices. If we replay a scenario over a long enough time frame, events will have happened which will not be directly included in the model, but could be implemented manually (e.g. introduce oil price spikes equivalent to those that happened in the real world). The better the scenario description, the easier to compare the effect in the model. Finally, even a close match between scenario and experimental results is no guarantee that we have correctly explained *why* this happened, only that we have explained something about how it could have happened. Corresponding results from a historic replay can increase confidence in the model, but non-corresponding results clearly indicate that the model has not captured enough of the underlying mechanisms and patterns (or at least not those that were important in the past).

3.10.2 Expert Validation

Expert validation is the most commonly used validation approach in agent-based modelling. Domain experts and problem owners (Balci 1994; Louie and Carley 2008) can discuss the behaviour of agents, the patterns of behaviour of system and also the application of model for its designed purposes. Modellers who have followed the steps presented in this chapter will have already have the conceptualisation validated due to high level through stakeholder engagement during the model building steps (cf. the "companion modelling" approach of Barreteau (2003)), but the outcomes can be validated by experts too.

Performing an expert validation usually includes structured interviews with individuals or workshops with larger groups of experts, systematically going through the model assumptions, mechanisms and outcomes. Although we should already have a good idea of which elements in the model cause these results (see Sect. 3.9), it might be wise to let the domain expert reason through the explanation on their own. Comparing the conclusions of the experts with those derived by the modeller will give a good indication if the model outcomes "make sense".

However, this approach is not without significant problems. Experts may have a good understanding of what *has happened* in the system in the past, but might not have a very systematic understanding on what *may* happen. Experts rely on their own internal models of system behaviour to estimate possible futures, all of which are subjective, biased and flawed in various degrees. Of course the developed model, also based on input from experts, will have the same biases, but these will be much more explicit than those of the models internal to an expert's head. A particularly dangerous trap is looking for experts to validate the model who already agree with the type of outcomes created, while ignoring experts with dissenting views. This is further exacerbated by dominant paradigms, which limit an expert's internal models to only those that are compatible with the general understanding of the state of the art. The solution might be that different experts, preferably experts with differing viewpoints or ascribing to slightly different paradigms, should give input to the model conceptualisation and decomposition stages as well as to the validation stages. Another pitfall is that experts make assumptions about how the model works, instead of really understanding it, which will be more likely the more complex the model is. It is essential to identify any experts' assumptions on the model and to encourage a focus on the observed model results rather than the possible limitations of the model.

Finally, a model validated in this way by consulting experts is often said to have *face validity*, which means the model "appears reasonable" and it "looks like" it will do what it is supposed to do. Experts can be wrong too, but expert validation remains an appropriate way to address the validation of agent-based models. In terms of practical application, models validated through experts can be considered good enough. If the decisions makers have had no access to a model they would have trusted the view of (probably the same) experts anyway.

3.10.3 Validation by Literature Comparison

Further validation can be performed by studying the academic literature. Similar conclusions reached through both theoretical research and non agent-based models increase the confidence in the model outcomes. When our model recommends a course of action compatible with an available theory or published case studies, we can claim an increase in model validity. Further validation is possible if other, non-agent-based, models of the same or similar systems problems exist, even if they have been used to answer a different question (Xu et al. 2003). Furthermore, it may be possible to gain further validity by performing new experiments that better align the assumptions, inputs and scenarios between the models as described in literature and those that seek validation. When performing such comparison experiments, we should not focus on replicating the exact outputs, but instead on the general outcomes and recommendations.

3.10.4 Validation by Model Replication

Finally, a strong—but labour intensive—form of model validation is model replication. Creating a second agent-based model with an different system decomposition, or a model using a different modelling technique (e.g. system dynamics) and comparing the outcomes can be a rich source of validation, and a good choice if the previously discussed options are unavailable. In an ideal situation a different research team will build these models, to remove any bias.

3.10.5 Step 9 Example

Our example model has received partial expert validation. The model is not based on real data, so could not be validated by comparison with real-world data through historic replays or literature evaluation. Designed as a proof of principle model, its aims were met when the problem owner expressed that such a model would indeed be valuable to them.

Expert Validation The model outcomes describing the spread of technologies through the social network of greenhouse farmers have been presented to a group of greenhouse farmers. The group clearly recognised the general dynamics of the technology diffusion outcomes and even suggested which individuals in the room might correspond to different abstract farmers in the model, especially related to the number of connections. The experts also immediately recognised the limitations of the model and identified aspects that would need to be added, such as the capital markets that enable technology adoption. In this case the expert validation is not a matter of concluding that this model faithfully corresponds to reality, but the process rather provides the modellers with insight into which aspects and outcomes are "good enough" and which need extra work to provide a useful model to the problem owner.

3.11 Step 10: Model Use

We opened this chapter by stating that the modelling process starts with a problem that requires solving. This problem generally takes the shape of an absence of knowledge about a real-world system or the response of such a system to possible interventions. The main advantage of agent-based modelling is the ability to explore the problem, look critically at what might be in that knowledge gap, and provide an incomplete range of possible system futures. Having arrived at the last step in the modelling cycle, we can start to use the results of the previous steps to hopefully rectify this lack of knowledge. In this section we explore some of the practical aspects of using models to answer the particular problem that a problem owner might have.

3.11.1 Outcome Presentation

Stakeholders are not often modellers themselves and may not even be computer literate or familiar with scientific methods and processes. Non-modellers are less likely to understand the limitations of model design, performance and outcome. They are rarely interested in the intricacies of the model or the experimental setup, and may be difficult to present results to if the presentations are not properly addressed.

Usually stakeholders look for a clear and simple answer to the question which started the whole 10 step process, perhaps even pushing for a "yes/no" answer or a single "optimal" value for a certain parameter at a certain point in time. Modellers must therefore take great care when presenting results to correctly portray the often subtle meaning of the results, and accompany those results with an intelligent amount of context to avoid misconstrued or applied badly conclusions. Assumptions about data or behaviour must be made explicit, as well as any limitations of the model, the experiments or the results. Modellers must resolve to present the results as simply and clearly as possible, but no more simply than is accurate just to bow to pressure for a short and sweet answer from busy problem owners, who need to draw their own conclusion based on these results and to take the responsibility for that conclusion.

Advances in computer graphics and data presentation allow for visually impressive images (see also Sect. 3.9.2.2). To those less familiar with computers, models or graphic heavy presentations tend to be overwhelmed by the amount of information presented or misled by visually pleasing images that are valued more than "boring" but more informative graphs. This may end in one of two disastrous ways: the information overload causes the audience to shut off and discard the entire model and result set as too complicated to use, or, possibly even worse, mentally reduce the information to a few take-home points that are accepted as "the truth". Furthermore, stakeholders tend to be far less critical of "pretty" model outcomes than less attractively presented ones, and of outcomes that support a conclusion already held by the problem owner than those that endorse an opposing standpoint or course of action. A thorough explanation of the model limitations should accompany any pretty results to prevent any inappropriate enthusiasm and special care should be taken to avoid cherry-picking the small subset of supporting results from a mixed collection of data.

To summarise, modellers must be sure that the presented information has clear, memorable take home points to prevent information overload, but that these take home points are exactly the ones intended and that they are not divorced from the context of the model and modelling process.

3.11.2 Raising New Questions

Modelling is iterative. As much as modellers might like to think that research can provide the definitive answer to a question, there will always be more new questions raised than answers found. These may be completely new questions discovered through the modelling process, or they may be issues not adequately covered by the model, but which the experiments show warrant further investigation. Perhaps a mechanism of the model can be improved through better assumptions, new experimental conditions are considered for examination, or the model might be extended by including more complexity. The problem owner will have learned something new after the outcome presentation and may now express a desire to explore other parts of the system. These new questions could motivate a new modelling cycle, starting again at Step 1 by defining a new problem and new relevant actor roles, taking on board the insights gained from the exercise.

Deciding whether to continue with a model to answer additional questions can be a delicate matter of balancing several conflicting objectives. Modellers from academia maybe only want to continue if the questions raised are scientifically interesting, or they may want to consider providing additional work in return for funding of other research projects. For any given model, there is likely to be a combination of both motivations, but modellers must be clear on what the balance between them is in order to remain objective about question raised, assumptions made and effort required. Modellers working as consultants from the private sector may want to extend any project if the stakeholders remain willing to pay for additional model development and advice, but even then it is important to determine the best interest of the client so that a balanced decision can be made about the future of the project.

3.11.3 Long Term Stakeholder Engagement

A model is usually too complex to just hand over to non-modelling stakeholders. Instead of just delivering "the model" as a final product to the stakeholder, modellers need to realise that the actual product is the insight into the stakeholders problem. In our experience it is beneficial to develop a relationship with the stakeholders and aid them in interpretation and presentation of the model and its results. Long term, iterative modelling exercises can be beneficial for both sides. Stakeholders receive advice and insight into the real problem at hand instead of problematic software or voluminous results without context and modellers, in return, develop a better understanding of the real issues and a increased motivation for taking responsibility for the application of the model.

3.11.4 Agent-Based Models and Stakeholders

Humans are relatively weak at systematic reasoning across myriad interactions between system elements, but we excel at pattern recognition. Computers, on the other hand, can easily systematically exercise the relations between system elements, creating complex behavioural patterns as output for us to interpret. So agent-based models, built on both objective system elements as well as the beliefs and assumptions of stakeholders, work through the innumerable actions of these system elements. As a result, our natural pattern recognition abilities, freed from the tedium of calculations, can formally test the otherwise untestable understanding and intuitions of the stakeholders, as well as delight in the emergent properties—the logical but nonetheless surprising patterns—caused by unpredicted actions and interactions between agents.

Our experience with stakeholders, users and modellers teaches us that agents (as well as their behaviours and their interactions) are easily understood. Humans tend to think about entities having goals and "wanting" things, so the narrative of agent actions and choices is generally intuitive, even though agent-based models offer a new modelling paradigm. While the agents may be intuitive, the emergence and self-organisation that arises from such a bottom-up, distributed approach is not. Those not used to modelling, or more familiar with a traditional top-down, command and control paradigm, may be distrustful of conclusions drawn from agent-based modelling results.

In fact, being a new modelling paradigm can be seen as an advantage in its own right. The new is often perceived as exciting, with modellers and users alike drawn to "shiny new toys" like moths to the irresistible glow of an electric lamp on a dark night. Using cutting edge techniques and tools can increase the likelihood of a successful application or bids when the competition is slower on the uptake. However, modelling, agent-based or otherwise, is not a panacea, despite having been oversold to policy makers with the promise of a rational prediction of future system states. When some predictions turned out to be erroneous, it raised doubts about the usefulness of modelling in general. Users have to be educated about the scope and relevance of models, and particularly about what agent-based modelling results can be used for, while modellers need to be more careful about the power of the model results and predictions. The danger is no less great in cases where policy makers have accepted a model and are comfortable with its use, as they may use it in unintended ways and arrive at misguided conclusions.

3.11.5 Computer Models and Mental Models

Compared to what the models can do, there are of course infinitely more things they cannot do. There is a theoretical limit to any generative model of an evolving system because evolution is intractable. No model can reliably or exactly predict the future, but only provide good enough predictions based on the available knowledge and understanding. Models can help us to increase this knowledge and understanding, but they cannot replace it.

We are all guided by internal models of how the world works based on our own knowledge and understanding of everything we have experienced so far. We use them to predict what is likely to happen in the future, to understand what has happened in the past, or make decisions. These mental models are so pervasive and so essential to human cognition and behaviour that we are largely unaware of them and how they might differ between individuals, or between an individual and reality. It is fundamental to understand that neither the explicit nor implicit models of laymen or experts are accurate representations of reality. Although we might be tempted to rely on the models held by scientists, these are not necessarily free from inaccuracies, as shown in the relatively fast turnover in scientific paradigms and popular theories.

So model behaviour, either positive or negative, can be surprising if it exactly matches the results and expectations of the modellers and stakeholders, or if it strongly contradicts all the expectations. But this is a case of comparing the results of an explicit, agent-based model to the implicit predictions of mental model, neither of which can really be said to be accurate. The difficulty here lies in examining the logic of both models, carefully scrutinising the reasoning and assumptions guiding them that have led to the unanticipated results. While stakeholders might be content with surprisingly positive results, the modeller must take special care to be sure that the results are not merely an artifact of unrealistic assumptions or model simplifications. Having said that; counterintuitive results are not always wrong either and it may be the mental models that need an adjustment.

As such, good and helpful results may be ignored if the mental models of those in the field are too restrictive to allow for any adjustments. While accidental mis-use is a problem, all of science is also subject to deliberate mis-use as well. Modellers must be wary of self interested groups that might misuse data, for example by cherry picking, suppressing, or falsifying data, or by using tactics to misdirect or detract from the results. Particularly problematic, modelling results may be subject to attacks that play up emotional responses to the conclusions rather than paying attention to the data. People feel strongly about issues, but they may not be able digest the specifics of the model or the output in order to include the conclusions in a reasonable argument.

3.11.6 Step 10 Example

The greenhouse model was first presented at a meeting of an association of greenhouse growers, where they found the model useful, interesting and reasonable. As this represented an early milestone in the development of a model that sought to explain a real-world phenomenon, this was enough validation. But this presentation also brought up aspects of model use. For example, because the presentation was fairly short, only taking up a limited time in an otherwise full schedule at the growers' association meeting, it is likely that the model, the interactions that generated the results, and recommendations made as a result were not fully understood by all of the attendees. For example, when presenting the social network graph, as grower agents were added to the simulation, the graph layout algorithm would constantly rearrange the structure of the network, causing a meaningless jiggle. The stakeholders were visibly impressed by this active graph display and repeatedly asked to the meaning of movement in the graph. While the motion is an, arguably visually appealing, artifact of the graphing algorithm, it did severely distract the stakeholder and impeded their understanding of the outcomes. Many might have agreed that the model was valuable purely as a consequence of a quality presentation, or of a reliance on science and modelling to uncover useful information, or they may have felt that they understood the model through sympathetic anthropomorphising of the agents.

Possible misunderstandings at this point are less consequential than they might have been. The model will continue development before the results are used to guide any major decisions. The growers will continue to be involved, helping to pin down aspects that were not well captured, refining unsubtle assumptions, and elaborating on the interactions. This early presentation as part of a longer term engagement prevents the modellers from walking into unrealistic dead end models, and ensures that even if the full value of a model was not understood by all audience members in the initial presentation, motivated participants will have the opportunity to express any concerns and ask new questions as the work continues.

It cannot be stressed enough that modelling is most effective when iterative. The cycle of asking questions, formalising ideas, implementing them, experimenting, and presenting results cannot be seen as a single indivisible pathway, but instead, perhaps it is more like a game of "snakes and ladders" where every turn attempts to move forward, but forward progress may involve repeating some steps many times.

3.12 Chapter Conclusions

To summarise, the development of agent-based models requires a careful analysis of the problem and the identification of the various actors involved. The system under study can then be decomposed into the elements which need to be modelled, a process which is followed by the formalisation of these concepts. Next, the behaviour of the active elements in the system (i.e. the agents) is mapped and described first as a narrative and then in pseudo-code. At that stage the modeller can start the software implementation, making use of existing tools (including those for software development, knowledge representation and knowledge management) and possibly existing model elements. With an implemented software model, the verification stage can start in which it is analysed if the model is implemented correctly. Before the model can be used, the experimental setup needs to be prepared and analysis of data is to be considered. Finally, after validating the model against the real system and the specific problem in mind, the model results can be used to support the problem owner.

This chapter presented the 10 steps in the modelling process of building agent-based models of socio-technical systems. In Part II of this book these steps are repeated for a number of case studies, and for each case study a detailed description of what was done during each step is given. With the practical guidelines from this chapter combined with several case-specific implementations, the reader should be able to start designing and implementing an agent-based model of a new socio-technical system.

Acknowledgements The authors would like to express their gratitude to all the other authors of this book for their input on the modelling practice.

References

Abela, A. (2008). *Advanced presentations by design: creating communication that drives action*. Pfeiffer.

Abela, A. (2011). Identify the most effective graphical elements to use in your presentation. http://www.extremepresentation.com/design/charts/.

Aldea, A., Bañares Alcántara, R., Jiménez, L., Moreno, A., Martínez, J., & Riaño, D. (2004). The scope of application of multi-agent systems in the process industry: three case studies. *Expert Systems with Applications*, *26*(1), 39–47.

Balci, O. (1994). Validation, verification, and testing techniques throughout the life cycle of a simulation study. *Annals of Operations Research*, *53*(1), 121–173.

Barreteau, O. (2003). Our companion modelling approach. *Journal of Artificial Societies and Social Simulation*, *6*(1).

Bluman, A. (2004). *Elementary statistics*. Boston: McGraw-Hill.

Epstein, J. (1999). Agent-based computational models and generative social science. *Complexity*, *4*(5), 41–60.

Epstein, J. M. (2008). Why model? *Journal of Artificial Societies and Social Simulation (JASSS)*, *11*(4), 12.

Gennari, J. H., Musen, M. A., Fergerson, R. W., Grosso, W. E., Crubezy, M., Eriksson, H., Noy, N. F., & Tu, S. W. (2003). The evolution of Protege: an environment for knowledge-based systems development. *International Journal of Human-Computer Studies*, *58*(1), 89–123.

Gruber, T. R. (1993). A translation approach to portable ontology specifications. *Knowledge Acquisition*, *5*(2), 199–220.

Hamming, R. (1962). *Numerical methods for scientists and engineers*. Columbus: McGraw-Hill.

Kasmire, J., Nikolic, I., van den Berg, W., & Bergwerff, L. (2011). Universal darwinism in greenhouses: proof of concept agent based model. In *ICNSC '11 proceedings of the 2011 IEEE international conference on networking, sensing and control* (pp. 56–61). New York: IEEE.

Louie, M. A., & Carley, K. M. (2008). Balancing the criticisms: validating multi-agent models of social systems. *Simulation Modelling Practice and Theory*, *16*(2), 242–256.

Matsumoto, M., & Nishimura, T. (1998). Mersenne twister: a 623-dimensionally equidistributed uniform pseudo-random number generator. *ACM Transactions on Modeling and Computer Simulation*, *8*, 3–30.

McConnell, S., & Books (1993). *Code complete: a practical handbook of software construction* (Vol. 1). Microsoft Press.

Newman, M. (2003). The structure and function of complex networks. *SIAM Review*, *45*, 167.

Nikolai, C., & Madey, G. (2009). Tools of the trade: a survey of various agent based modeling platforms. *Journal of Artificial Societies and Social Simulation*, *12*(2).

Nikolic, I. (2009). *Co-evolutionary method for modelling large scale socio-technical systems evolution*. PhD thesis, Delft University of Technology.

North, M., Collier, N., & Vos, J. (2006). Experiences creating three implementations of the Repast agent modeling toolkit. *ACM Transactions on Modeling and Computer Simulation*, *16*(1), 1–25.

Noy, N. F., & McGuinness, D. L. (2001). *Ontology development 101: a guide to creating your first ontology*. Technical Report SMI-2001-0880, SMI. http://protege.stanford.edu/publications/ontology_development/ontology101-noy-mcguinness.html.

Robert, C. (2004). *Monte Carlo statistical methods*. Berlin: Springer.

Sargent, R. G. (1998). Verification and validation of simulation models. In *Proceedings of the 30th conference on winter simulation*, Washington, D.C., United States.

Tang, B. (1993). Orthogonal array-based Latin hypercubes. *Journal of the American Statistical Association*, *88*(424), 1392–1397.

Wang, H. H., Noy, N., Rector, A., Musen, M., Redmond, T., Rubin, D., Tu, S., Tudorache, T., Drummond, N., Horridge, M., & Seidenberg, J. (2006). Frames and OWL side by side. In *9th international protégé conference*, Stanford, California, USA.

Wu, C., Hamada, M., & Wu, C. (2000). *Experiments: planning, analysis, and parameter design optimization*. New York: Wiley.

Xu, J., Gao, Y., & Madey, G. (2003). A docking experiment: swarm and repast for social network modeling. In *Seventh annual swarm researchers conference (Swarm2003)*, Notre Dame, IN.

Ye, K. (1998). Orthogonal column Latin hypercubes and their application in computer experiments. *Journal of the American Statistical Association*, *93*, 1430–1439.

Part II
Case Studies

Chapter 4
Introduction to the Case Studies

K.H. van Dam, I. Nikolic, and Z. Lukszo

Abstract In the first chapter of Part II of this book an introduction to the case studies presented in the following chapters is given. The theory and practice presented in Chaps. 2 and 3 are applied to the development of models of various socio-technical systems. Three different time scales for the case studies can be identified: short term (*operational*), middle term (*tactical*) and long term (*strategic*). The first case study deals with operational decision making in supply chains, in which abnormal situations can disrupt the normal operation guided by the actors in the system. In the second case study a longer time period is considered, and tactical decisions about consumer behaviour regarding lighting is studied. Case studies three and four deal with the long term. Case study three looks at the strategic decision making of electricity producers, including the effects of several policies and carbon taxation or carbon emission trading schemes on the portfolio development. In the fourth case study the life cycle of mobile phones is considered, with emphasis on the recycling of metals used in their production. Together, these case studies give an overview of the possibilities of agent-based models for socio-technical systems. All models in this book are based on the same shared ontology for socio-technical systems, which describes key concepts for networks of actors as well as technical components, along with a detailed set of properties for the physical and economical aspects of the system. The basic structure of this ontology is described here and the workflow using the ontology is explained.

4.1 Case Studies

Part I of this book concentrated on presenting the theory and practice of agent-based model development for decision support in socio-technical systems. Chapters 2 and 3 showed which theories from complexity sciences, large scale socio-technical systems and computer science are relevant for analysing these systems and how this is reflected in the challenge of developing a computational model, and

K.H. van Dam (✉)
http://www.koenvandam.com
e-mail: k.van-dam@imperial.ac.uk

K.H. van Dam et al. (eds.), *Agent-Based Modelling of Socio-Technical Systems*, Agent-Based Social Systems 9,
DOI 10.1007/978-94-007-4933-7_4, © Springer Science+Business Media Dordrecht 2013

which steps have to be taken from the problem formulation to using the model. In Part II this knowledge is applied to the development of five different models. The various case studies have all been developed by different teams of modellers, who each have their own take on the theory and practical steps, and have used them to their best advantage for the domain they were addressing. The cases show how the steps lead to models that are useful in a wide range of domains and policy issues, illustrating the wider applicability of the approach.

Three different time scales for the case studies can be identified: short term (*operational*), middle term (*tactical*) and long term (*strategic*). The following sections briefly introduce the models for each of these time scales. Together, these models give an overview of the possibilities of agent-based models for socio-technical systems and the challenges modellers encounter when dealing with the particular aspects of the domain or the problem, while at the same time serving as applications of the theory and practice of Part I, hopefully leading to increased insight into the modelling trajectory.

4.1.1 Operational Models

The first case study (Chap. 5) deals with operational decision making in supply chains, in which abnormal situations can disrupt the normal operation guided by the actors in the system. New decisions have to be made to best respond to an unexpected situation in which a certain event triggers the scheduled activity to be adjusted. Two models of two different supply chains, namely an oil refinery and a multi-plant chemical enterprise, are presented to give insight into how agent-based simulation models can be employed for decision support at the operational level.

4.1.2 Tactical Models

In the second case study (Chap. 6) a longer time period is considered, and tactical decisions about consumer behaviour regarding lighting are studied. Household consumers have a wide choice of available lighting options, including several variations of energy saving bulbs, and their behaviour highly depends on the behaviour of others within their social network. Different scenarios of policies by the government have to be evaluated to see how they will influence the energy use and carbon emissions related to consumer lighting. The agent-based model developed in this chapter can be used to test the effectiveness of such policies.

4.1.3 Strategic Models

Case studies three and four deal with the long term. Case study number three (Chap. 7) studies the strategic decision making of electricity producers, including the effects of several policies and carbon taxation or carbon emission trading schemes on the portfolio development. The model includes agents representing power production companies which own, operate and invest in power plants while they sell electricity and buy fuels as well as emission rights. The consequences of policies on long term developments of the whole system can then be analysed with this model.

In the fourth case study (Chap. 8) the life cycle of mobile phones is considered, with emphasis on the recycling of metals used in their production. The simulation model can be used to evaluate which changes in the behaviour of the stakeholders or in the environment have a positive effect on closing the loops. Actors in the mobile phone life-cycle are modelled as agents, including manufacturers, consumers, refurbishers and recyclers which trade metals, mobile phones and mobile phone components. The model provides insight into the dynamics of global flows as a consequence of changes in the behaviour of actors and in the properties of physical goods.

4.1.4 Setup of Case Study Chapters

To guide the reader through several examples of model development, all case studies are structured following the modelling steps explained in Chap. 3. Each chapter in Part II follows these same ten steps, making clear how the model was developed, how experiments were performed, and how the lessons learned have been applied to help solve a specific problem by supporting decisions made by the problem owner. All chapters use the following section headings:

- Introduction
- Step 1: Problem formulation and actor identification
- Step 2: System identification and decomposition
- Step 3: Concept formalisation
- Step 4: Model formalisation
- Step 5: Software implementation
- Step 6: Model verification
- Step 7: Experimentation
- Step 8: Data analysis
- Step 9: Model validation
- Step 10: Model use
- Conclusions

This means that for each model development step the relevant section from Chap. 3 can be used as a quick reference and that the results of the steps can directly be compared between the case studies presented. This should also give the reader an impression of the different applications and the width of the solutions available to tackle specific modelling needs.

4.1.5 State-of-the-Art Modelling

Following the case studies in these three time scales and following the ten steps, a possible future direction for modelling socio-technical systems is proposed in Chap. 9, focusing on collaborative modelling and open data. The mobile phone recycling case is used to demonstrate the latest developments for using ontologies, real data, and advanced agent behaviour. This case study pulls together all the work performed for the various other projects and moves it forward by introducing the latest insights from modelling, artificial intelligence and semantic technologies. It may not be the starting point for a new modeller, but serves to highlight the possibilities of advanced tools and elaborate simulations. Combined, the case studies in Part II showcase how the theory and practice from Part I can be applied to tackle real problems in socio-technical systems.

4.2 An Ontology for Socio-technical Systems

All case studies in this book are based on the same shared ontology for socio-technical systems, which describes key concepts for networks of actors as well as technical components, along with a detailed set of properties for the physical and economical aspects of the system. To support the reader in understanding the concept formalisation used in these models, the basic structure of this ontology is described here and the workflow using the ontology is explained. Following the introduction of the concept of an ontology in Sect. 3.4.2, in which an ontology was defined as "a formalisation of a conceptualisation" (Gruber 1993), this chapter thus gives an example of an ontology developed specifically for modelling socio-technical systems using agent-based models.

The ontology shown here was the result of studying the theory as discussed in Chap. 2 and going through the practical steps from Chap. 3 over several iterations while building models, among which the ones used in this book. Ontology design is very much linked with the aims the designer had in mind as well as the specific view of the world that is used, so it is not the intention to present the *only* ontology to be used for all socio-technical systems, but rather to give an example of a formalisation which has been proven to be applicable and useful.

4.2.1 Aims

The goals of developing a shared ontology for this class of systems can be summed up as *interoperability* and *inter-connectivity*. Because socio-technical systems involve a mixture of social and physical elements which interact in various ways and the domains are often highly interlinked, a shared formal language was needed to make sure models and model components could work together (allowing interoperability) and to ensure that models of different domains could be put together if required (providing inter-connectivity). The approach was based on modularity and the definition of "building blocks". The functional requirements can then be summed up as follows. The ontology should:

- support a wide range of socio-technical infrastructure systems including industrial clusters, supply chains, energy and transport networks;
- offer flexibility for experiments with varying configurations of the social and technical networks;
- be fully modular;
- be easy to use by modellers, including those not involved in the development of the framework;
- extendible so that case-specific aspects can be added; and finally
- be easy to explain to new modellers, but especially to the problem owner, stakeholders and domain experts.

4.2.2 Development

The development of the ontology as presented here was an iterative process: by applying the framework and the ontology to case studies, changes were made that added to the shared concept formalisation. During each application cycle new building blocks were implemented and existing ones improved. It was necessarily teamwork, involving, among others, the authors of the chapters in Part II of this book. During the concept formalisation step (Sect. 3.4) existing ontologies can be extended or refined with new abstract classes and additional properties. Some new additions were made specifically for certain cases, but others were more widely applicable and could be shared and fed back into the framework. As new "building blocks" became available, more and more could be re-used. For some models it might even sufficient to *only* use existing concepts and building blocks, with minor case-specific modifications.[1]

While the ontology was maturing through application after application, not only were new elements added to the framework, but also old ones were modified by adding properties or completely redefining concepts, for example by changing the hierarchy and inheritance of classes. Especially during the initial phase such redefining of concepts was often needed as only through application in models and by sharing with other modellers the framework could be shaped. Such significant structural changes meant that earlier models were no longer compatible with the new ontology and had to be adjusted. This was part of a steep learning curve. The challenge was then to demonstrate that such major revisions are no longer needed to model socio-technical systems and that indeed only small additions are sufficient

[1]The models described in this book are all of such complexity and on new territory that they all required specific concepts to be newly formalised in the ontology.

for new cases. By analysing all revisions[2] made to the ontology over time it was demonstrated that indeed the ontology can be considered to be complete (i.e. suitable for the various domains that fall within the scope of socio-technical systems and new models are built based on already defined concepts), correct (i.e. no major repairs are required) and usable (i.e. useful also for people who have not been involved in the initial development) (van Dam 2009).

4.2.3 Key Concepts

In this section some of the key concepts as formalised in the ontology for socio-technical systems are introduced. This is not an attempt to provide a complete picture, especially since other applications may require their own ontology. However, because all case studies in this book use this ontology, a broad overview of how it has been set up and structured is beneficial to better understand the model developments in rest of Part II. For a complete picture the reader is referred to van Dam (2009) and for a discussion on the *social process* that enables the development of such an ontology please see Nikolic (2009).

Figure 4.1 shows a fragment of the ontology with the relationship between different classes of Nodes (i.e. Social and Physical) and some of their properties. Agents and Technologies are both nodes that, together, form the socio-technical network. Classes inherit properties from their superclass through the 'is a' relationship where the properties can be seen as 'has a' relationships.

All nodes share certain properties, such as the fact that they have a label, a set of physical- and economical properties and a number of "edges" defining in- and outgoing links with other nodes. Using this definition, an Agent, representing actors in the real world, potentially 'has a' Technology (or possibly more than one) which it owns, maintains or controls, for example. The Technology has specific properties allowing the definition of how the technology works, what it produces and what infeeds it requires, among other things.

Figure 4.2 describes the concept of an OperationalConfiguration, which is a "recipe" of inputs and outputs for this particular unit. For each input or output, through a ComponentTuple, the name of the product, the amounts, the unit in which the amount is specified, and several properties and labels can be defined. This provides a rich language with which technologies of any scale can be described, from a small light bulb to a large oil refinery, as will become clear in the case studies.

As a last example of a key concept in this ontology, Fig. 4.3 shows the connections that can exist between one PhysicalNode (such as a Technology, c.f. Fig. 4.1) and another one. A PhysicalConnection represents a physical link, such as a pipe or a cable, through which goods can be transported resulting in a PhysicalFlow in which mass is transferred between nodes. Similarly to the physical edges, social

[2]Having all previous versions of the ontology available for such analysis is yet another benefit of using version control (see Sect. 3.6.2.1).

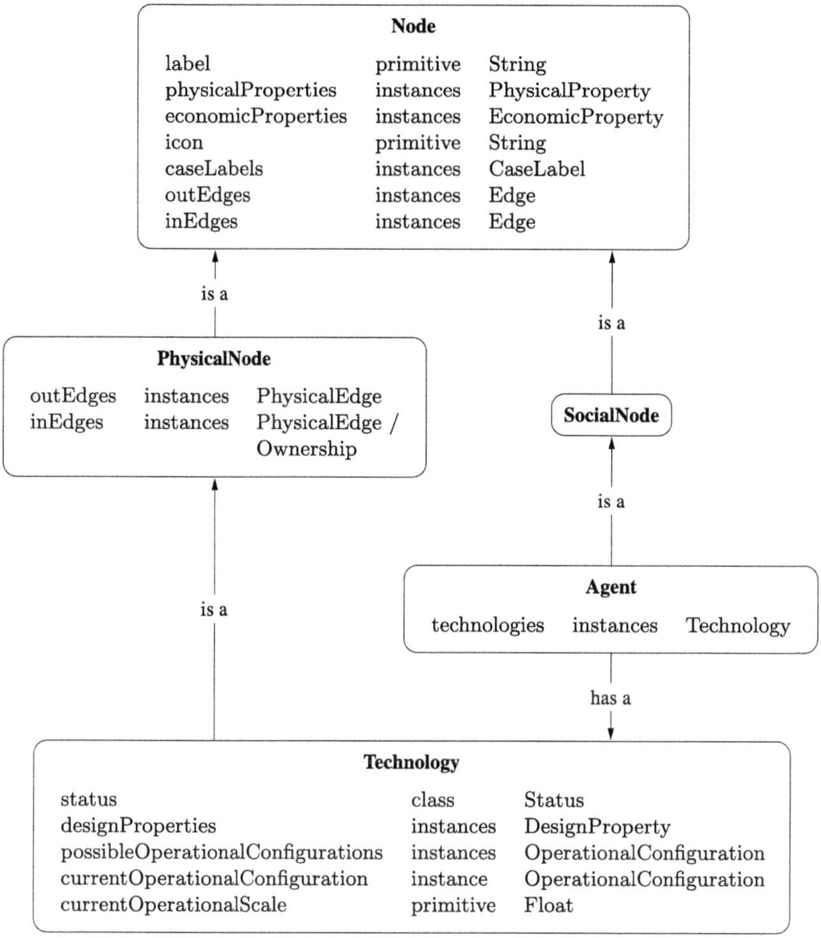

Fig. 4.1 A fragment of the ontology for socio-technical systems, showing the relationship between different classes of Nodes (Social and Physical) and some of their properties

edges including concepts such as contracts can be defined to describe in great detail the communication between agents about trading a material flows and any arrangements for transporting them. The links between agents form a dynamic social network, while the results of these negotiations is a (possibly ever-changing) physical network. These layers together then represent the social-technical system which has to be modelled.

4.2.4 Usage

This ontology is used in the modelling process in two ways. First, it serves as a structure for a *knowledge base* containing entities, from actors to technologies, which

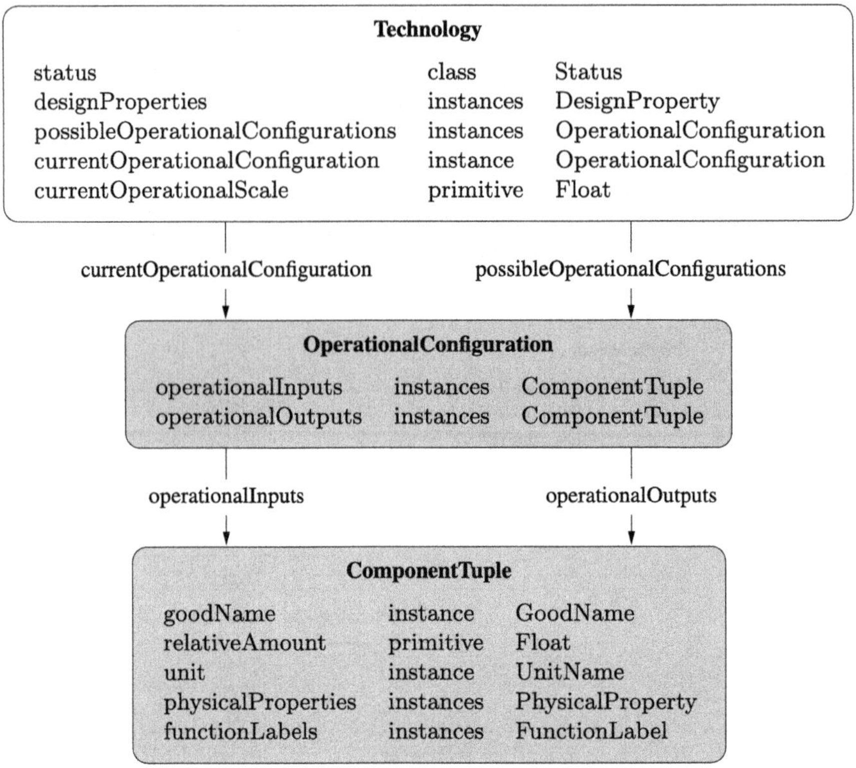

Fig. 4.2 OperationalConfigurations for the class Technology

play a role in the models. A shared knowledge base allows exchange of these entities as well as re-use in various models. The modeller can add new instances or base the model on existing descriptions, or a combination. Once the instances have been defined through a graphical user interface and stored in a central location, the agent-based model can be initialised by reading this information and instantiating the agents and their properties in the simulation.

Secondly, the concepts as formalised in this ontology form the language in which to encode the behaviour and the communication of the agents. The ontology makes it possible to define, for example, how an agent chooses the best contract for a deal and which properties the traded goods have. When multiple agents are defined using the same ontology, *interaction* between them becomes meaningful because it is confirmed that words that are part of the communication have the same corresponding intentions. The modeller can base algorithms (and even pseudo-code in Step 3, see for example Algorithm 5.1) on the words as defined in the ontology, as will become clear in the chapters presented next.

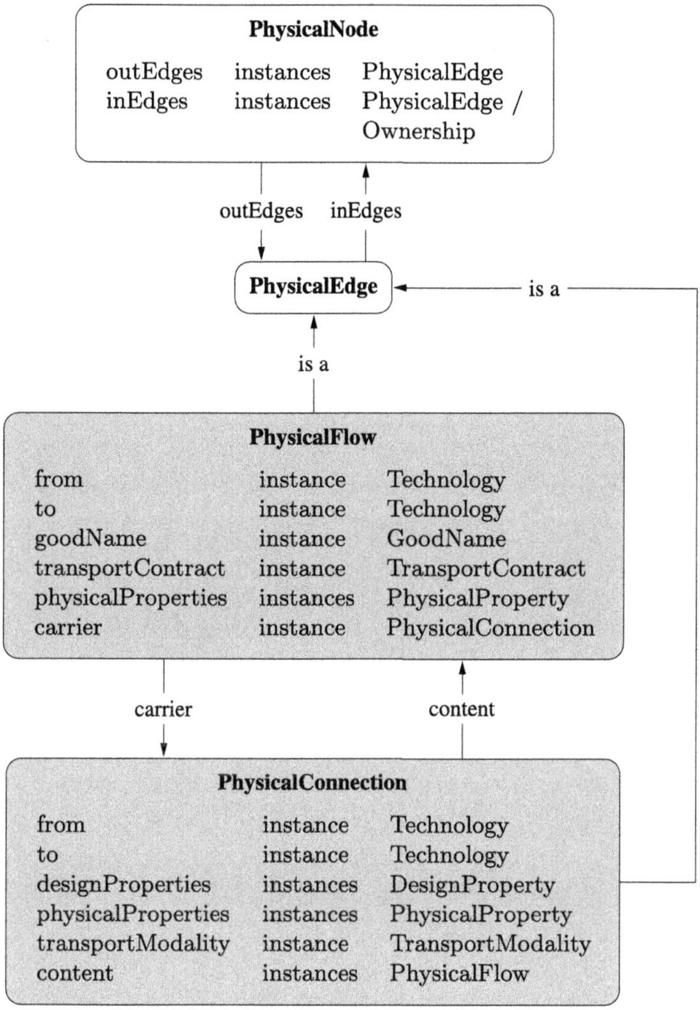

Fig. 4.3 Physical edges in the ontology

References

Gruber, T. R. (1993). A translation approach to portable ontology specifications. *Knowledge Acquisition*, *5*(2), 199–220.

Nikolic, I. (2009). *Co-evolutionary method for modelling large scale socio-technical systems evolution*. PhD thesis, Delft University of Technology.

van Dam, K. H. (2009). *Capturing socio-technical systems with agent-based modelling*. PhD thesis, Delft University of Technology, Delft, the Netherlands. ISBN 978-90-79787-12-8.

Chapter 5
Agent-Based Models of Supply Chains

B. Behdani, K.H. van Dam, and Z. Lukszo

Abstract Based on the modelling steps discussed in Part I, this chapter aims to present ways in which agent-based simulation models of supply chains can be developed and used to improve the performance of these systems in both normal and abnormal situations. An industrial supply chain with a network of several independent companies is a good example of a socio-technical system. The physical and social networks of the actors involved in their operation collectively form an interconnected, complex system in which the actors determine the development and operation of the physical network and, likewise, the physical network affects the behaviour of the actors. In this type of system, the many interactions taking place in the social and physical subsystems can result in the complex, dynamic behaviour of the supply chain as a whole. Accordingly, any attempt to improve the functioning of the supply chain requires a comprehensive understanding of this behaviour under different supply network configurations. Most of the current approaches to the modelling and simulation of supply chains do not capture the rich socio-technical dynamics present. The agent-based modelling approach, however, seems to be very promising as a means to address this complex behaviour. To demonstrate its applicability, we will present agent-based simulation models for two different industrial supply chains: an oil refinery and a multi-plant chemical enterprise. Using the models described in this chapter, the outcomes of the system under a broad range of possible agent behavioural rules and environmental events can be explored, and improved levels of system functioning can be identified.

5.1 Introduction

Supply chains present an interesting research subject from a socio-technical systems perspective. In these chains, complex production technologies interact with geographically distributed, intelligent, autonomous entities; each with its own dynamics, goals, desires and plans. There is a significant challenge in modelling such systems which function in dynamic, stochastic, socio-economic environments and

B. Behdani (✉)
Delft, the Netherlands
e-mail: b.behdani@tudelft.nl

K.H. van Dam et al. (eds.), *Agent-Based Modelling of Socio-Technical Systems*, Agent-Based Social Systems 9,
DOI 10.1007/978-94-007-4933-7_5, © Springer Science+Business Media Dordrecht 2013

show intra-organisational and inter-organisational complexity. The focus in this chapter is on *operational decision-making*, which has a relatively short time frame and in which decisions and actions are directly linked to the functioning of production units and the transportation of mass flows.

This chapter presents two case studies: one is a project designed to help an oil refinery in dealing with disturbances and ensuring smooth operation at minimal cost; the other is support for a multi-plant chemical enterprise, aimed at helping them to respond to different types of abnormal situations, including unexpected plant shutdown. The two systems lie in a similar domain (i.e. supply chains) but their individual characteristics and the specific problem formulations result in different models that are used in different ways. The models have a common starting point, but these diverge into different paths because they address different problems and thus have different conceptualisations and implementations. This chapter will demonstrate the link between the problem and domain characteristics of the two cases and the choices made during the development and use of the models. The cases illustrate the possibilities offered by agent-based models in the field of supply chain management. Furthermore, the description of the modelling choices for these case studies demonstrates how agent models of such complex systems can be developed in practice.

5.1.1 Supply Chains

A set of companies whose activities can together be characterised as the production, service provision, distribution or trade of raw materials, by-products or market goods is regarded as an "industrial network". A supply chain is a good example of such a network. Along each *supply chain*, numerous links and nodes exist, grouping material suppliers, production facilities, distribution services, warehouses, service centres and customers into a single cohesive entity. The *actors* in the supply chain coordinate their activities and collaborate together to improve the efficiency of the whole system in general and their own enterprise in particular. The *physical* elements of the supply chain encompass processing equipment which may be located at different sites. As the complexity of this system increases, the likelihood of a breakdown at any point in the network increases as well. This can happen, for example, due to the heavy interdependence of the network components, lengthened supply and distribution lines or heightened competition on the market. As a consequence, different independent decision-makers in the supply chain might come into conflict; something which can lead to the instability of the system and place the network in an abnormal situation.

5.1.2 Abnormal Situation Management

Abnormal situations in supply chains encompass a range of events outside the "normal" operating modes, including human error, fires, shipment delays, (unplanned) maintenance and equipment failure. Note that disturbances may be part of normal operation. When the normal operating mode is interrupted by a disturbance or series of disturbances, such that the network cannot function properly, this is called an *abnormal situation*. In such a case, the planned production targets may not be met unless swift action is taken to minimise the negative effects. Abnormal situations are rarely local; a disruption in one part of an industrial network can influence the performance of other components and the network as a whole. This raises the following questions:

1. To what extent does one part of a network influence a system's overall performance?
2. What are suitable strategies and control mechanisms for coping with and recovering from abnormal situations?

This chapter describes how agent-based simulation models of supply chains are developed and used to support decision-makers in the management of abnormal situations in supply chains and to help answer these critical questions.

5.1.3 Supply Chain Modelling

Agent-based modelling is not the only modelling paradigm used for *supply chain modelling* and simulation. Over the past two decades, many other types of analytical models have been presented to study the dynamic behaviour of supply chains during normal and abnormal situations. Most of these models are based on mathematical programming and operations research approaches. Overviews of some key publications are given by Tayur et al. (1999) and Min and Zhou (2010). Although these analytical models cover a broad range of areas in supply chain management, from functional aspects (e.g. inventory management) to inter-organisational issues (e.g. collaboration along the supply chain), most of them do not explicitly take into account the social aspects of the system; hence, the dynamics of the modelled system are dominated by the physical realities rather than the interactions between decision-makers.

The socio-technical nature of supply chain problems, however, is the motivation behind using the agent-based modelling approach as an alternative paradigm. The benefit of this approach is that it takes an actor-centric perspective instead of modelling only the outcomes of their activities. The actions of each actor (represented as an agent) and the interactions between them are explicitly represented in such models, and consequently the behaviour of the entire system emerges (van Dam 2009). Swaminathan et al. (1998) provided one of the first efforts for modelling supply chains using agent-based modelling: a flexible, agent-oriented framework which enables rapid development and customised, agent-based decision support tools for supply chain management by using a library of structural and control elements (e.g. supply chain roles and policies). Other examples of using agent-based models for supply chains are given by Julka et al. (2002a, 2002b); Kaihara (2003); Siirola et al.

(2003) and the pioneering work of Gjerdrum et al. (2000). Moyaux et al. (2006) present a review of the applications of an agent-based paradigm for supply chain management. More recent work on modelling supply chains using the agent-based modelling approach includes Srinivasan et al. (2006), Mele et al. (2007), van Dam et al. (2008) and Behdani et al. (2010a, 2010b).

5.1.4 Chapter Organisation

The rest of this chapter will follow the ten model building steps, described in Chap. 3 of this book, for two case studies. The first two steps are discussed for the two models in Sects. 5.2 and 5.3, respectively. Afterwards, steps 3 through 10 are presented integrally in Sects. 5.4 through 5.11. In each of the steps the development of both models will be addressed, focusing on those aspects in which the development is influenced by specific choices stemming from different problem statements, application domains and the different conceptualisations that flow from this. The chapter thus does not provide a complete picture of the development of each of the models; instead it highlights the specific choices that were made in two related models and demonstrates where the models converge or diverge. In Sect. 5.12 these findings are summarised and concluding remarks on the lessons learned on agent-based modelling and supply chain management are given.

5.2 Oil Refinery Supply Chain

In this section the first two modelling steps ('Problem formulation and actor identification' and 'System identification and decomposition') are followed in a case study of an oil refinery supply chain.

5.2.1 Step 1: Problem Formulation and Actor Identification

A hierarchy of decisions has to be made in managing the supply chain of an oil refinery: strategic (e.g. capacity investments, the addition of units, technology upgrades and supply chain reconfiguration), tactical (e.g. production planning, policy evaluation, disruption management) and operational (e.g. procurement, storage, scheduling, throughput level) (Pitty et al. 2008). Disturbances in the supply chain, including equipment failures and shipping delays, have consequences for the operation of a plant. What is the best response to a certain situation, and which changes in operation or emergency procurement are required and efficient? During nominal process conditions the operation is optimised by choosing which crudes to buy, how much crude is needed and from which supplier to order. The mode of operation is scheduled based on predicted demands, and the throughput for operation of the refinery is set based on actual consumer demand. When an abnormal situation manifests itself, this normal approach is no longer adequate.

In the supply chain domain as considered in this case study, the following actors can be identified:

Problem owner: The supply chain manager is responsible for overseeing the operation of the entire supply chain. However, in some circumstances, the operator of a single node in the supply chain could be considered to be the problem owner.

Stakeholder: In addition to the supply chain manager, stakeholders include the managers of the various supply chain entities, shippers, consumers, local governments, national governments, investors, port authorities, etc.

Modeller/analyst: The authors of this chapter are together considered to be the modeller, whose role is to develop a model for the problem owner to help solve the problem as motivated above.

Facilitator: In this case the role of the facilitator is not applicable, because the model building process has been a more or less straight-forward exercise in a domain which has been well studied in literature. In future development and extensions of these models, a facilitator could be required to lead the discussions between stakeholders and modellers.

Domain expert: The domain expert for this case study includes literature sources in process systems engineering and computer-aided chemical engineering, as well as several experts in the academic field of supply chain research.

With this problem owner and these other actors in mind, the system is identified and broken down in the next section.

5.2.2 Step 2: System Identification and Decomposition

An *oil refinery supply chain* begins with the oil reservoirs, both onshore and offshore. Crude oil is tapped from these sites and then transported to various refineries around the world, mostly by pipelines or large ships (very large crude carriers, or VLCC). Transport times of crude are relatively long; it takes four to six weeks for a VLCC carrying crude oil from the Middle East to reach refineries in Asia, for example. The crudes are then processed in crude distillation units (CDU) and separated into fractions based on their boiling points. These fractions are processed further in different downstream refining units such as reformers, crackers and blending pools to arrive at the various products. A single crude mix may yield numerous products and their variants through a suitable alteration of processing conditions. Hence, refineries must adapt their operations to the different crude batches to meet the required product specifications.

The refinery occupies a pivotal position in the supply chain, as its functional departments initiate and control the interactions with the external entities, these being oil suppliers, third party logistics providers, shippers, jetty operators and customers. The other stakeholders as identified in Sect. 5.2.1, such as the national governments, investors and port authorities, are considered to be outside the scope of the model, given the problem statement. They may still play an important role in the interpretation of model results and they may (indirectly) be affected by the decisions made by the problem owner, but they do not have to be included as actors in the *model* of the system. The same holds true for the supply chain manager, who does not need to be included in the model itself. Even though the supply chain manager is a member of the refinery company, which is included as an actor in the model, his tasks and behaviour are not captured in the model because the model is not designed to offer *online* decision support. The supply chain manager is the *user* of the model and can choose and validate decisions while experimenting with different actions.

The operation of the refinery supply chain requires various decisions in every cycle: what mix of products to make, which crudes to purchase and in what quantities, which mix to process and in what processing mode, etc. Different actors are responsible for the different decisions (Julka et al. 2002b). These actors and their interactions are shown in Fig. 5.1. The entities, shown as blocks in Fig. 5.1, communicate with each other through information flows (dashed arrows) in order to control the material flows (solid arrows).

The refinery's physical units (shaded blocks) may be further sub-divided into storage units such as crude and product tanks and processing units such as the CDU, reformer, cracker and blending pools. The functioning of these units and other supply chain activities is overseen by the functional departments: the storage department and the operations department. The actors and physical systems are shown in Table 5.1. Note that some actors, for example the logistics department, do not own or control a physical system, but they do have their own specific tasks and communicate with the other actors. All actors in the model are briefly described below:

Refinery company — is the overall institution which owns the refinery. It can be broken down into various departments. Relevant properties include the company's fixed and liquid assets.

Operations department — is in charge of production and operates the oil refinery. It selects a mode of operation (i.e. a recipe) and the throughput of the refinery based on sales. Properties include the characteristics of the refinery units.

Storage department — is responsible for managing the storage tanks for crudes and end products and monitors levels to ensure that limits are not violated. The storage department also alerts the operations department if there is not enough crude in storage during operation.

Sales department — is in contact with consumers about the sale of end products. This department also makes an estimate of

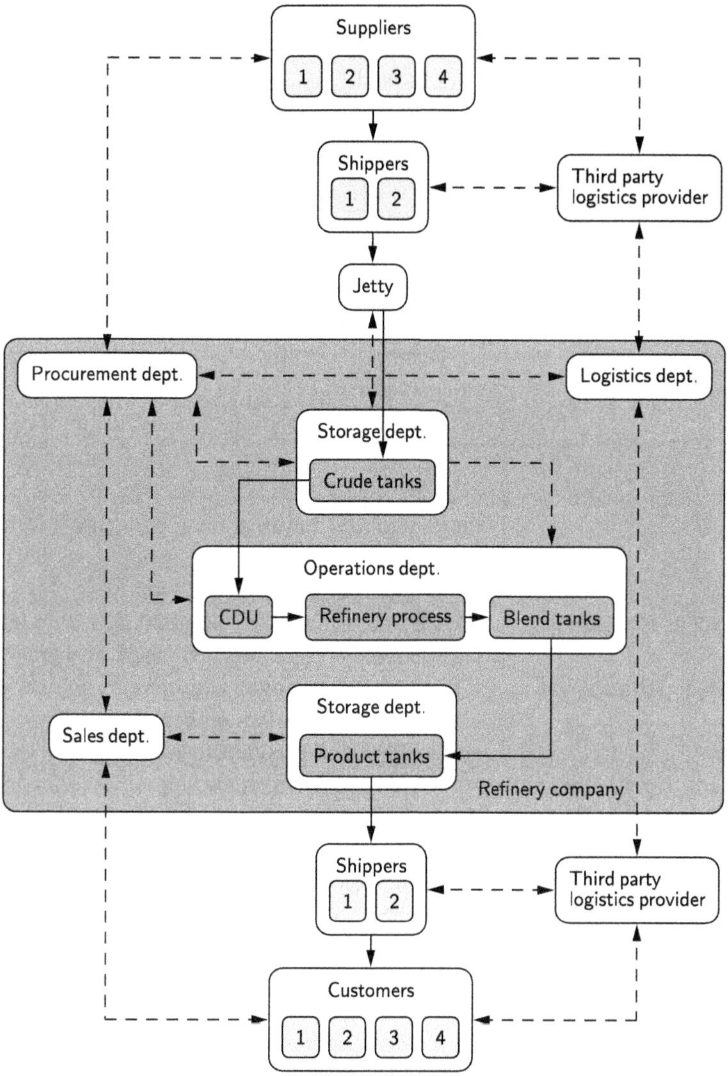

Fig. 5.1 Schematic depiction of an oil refinery supply chain (adapted from Pitty et al. 2008). Arrows with a solid line represent material flows; arrows with a dashed line represent information flows

Table 5.1 Agents and Physical Nodes and their relationships in the refinery supply chain case

Agent	Relationship	Physical Node
Refinery company	owns	Refinery units (incl. CDU)
Refinery company	owns	Raw materials storage tanks
Refinery company	owns	End product storage tanks
Operations dept. (Refinery)	controls	Refinery units (incl. CDU)
Storage dept. (Refinery)	controls	Raw materials storage tanks
Storage dept. (Refinery)	controls	End product storage tanks
Sales dept. (Refinery)	–	not applicable
Procurement dept. (Refinery)	–	not applicable
Logistics dept. (Refinery)	–	not applicable
3rd party logistics provider	–	not applicable
Shipper	–	not applicable
Supplier	owns	Oil wells
Jetty owner	owns	Jetty
Consumer	owns	Consumer installation

	the demand to be used for forecasting and later finalises the contracts for the actual sales. Properties include the prices of end products.
Procurement department	buys crude oil from the suppliers based on the forecast demand and the planned mode of operation.
Logistics department	is in charge of arranging shipping for the goods purchased by the procurement department as well as the shipment of end products to consumers.
Third party logistics provider	forms the link between the shippers and the logistics department.
Shipper	moves goods from one place to another.
Supplier	sells crude oils.
Jetty owner	is in charge of the jetty where crude oil tankers dock to unload the oil into the storage tanks. Furthermore, this actor manages the scheduling of vessels and operates the pump.
Consumers	have a demand for the end products of the refinery.

Other elements in the system are considered to be part of the "environment" of the model. This includes aspects which are assumed not to be influenced by components within the system. This means that, for example, the prices of crude oils and the demand for the various end products are assumed to be givens and are not generated within the model. Emissions during the supply chain activities (including production and transport) are not taken into account in this system description, as they are not relevant for the questions asked of this model.

5.3 Multi-plant Enterprise Supply Chain

In this section the first two modelling steps ('Problem formulation and actor identification' and 'System identification and decomposition') are followed for a case study of a multi-plant enterprise supply chain.

5.3.1 Step 1: Problem Formulation and Actor Identification

The *multi-plant enterprise* can be viewed as a *multi-level system*, whether hierarchically interconnected or decentralised, with a number of operational regimes at the various system levels. Usually, at each level of the decomposed system (i.e. functional level, plant level, enterprise level) local performance objectives are defined which should, preferably, not be restricted to the optimisation of local goals but rather aim at optimally contributing to the overall goal. However, the relation between local and overall system performance objectives may be rather fuzzy, especially since the overall objective is often not defined in detail and involves a longer time horizon. The local objectives are generally more detailed, concerned with a shorter time horizon and often with the specific interests of an individual actor (e.g. a business unit).

To facilitate the optimisation of the performance of the system as a whole, a coordinator may be required to supervise local decision-making in its relation to the overall goal. Therefore, a complex network of production plants needs a special form of operations management to coordinate plant-dependent activities and to ensure that the enterprise as a whole can realise its optimum performance (Bhatnagar et al. 1993). Operations management involves, among other things, decision-making regarding global enterprise resources planning systems: how to best operate a multi-plant enterprise and at the same time how to operate the separate plants in an optimal way.

These decisions can be categorised into long-, medium- and short-term decisions. The long-term decisions basically deal with operations strategy, product planning and the design of facilities and processes. The medium- and short-term decisions are concerned with the planning and control of production activities; more precisely, procurement and inventory management, sequencing and scheduling, quality management and maintenance management (Lee-Post and Chung 2008). Although some of those decisions can be made by an individual company, in a multi-plant enterprise each plant is a part of a corporate network and its relations with other plants are as important as its internal operations. With appropriate operations management, the enterprise can respond more quickly to dynamic market events, reduce its operating costs and increase customer satisfaction. Moreover, during an abnormal situation (e.g. unexpected shutdown of one production plant) it can be more resilient and recover from disruption more rapidly.

A multi-plant enterprise model can be helpful in finding a deeper view of network components' behaviours and their effects on the enterprise's overall performance.

Such a model can help decision-makers to study alternative scenarios and also to improve the operation with respect to some important performance indicators.

Even though the problem domain is different from the one discussed in Sect. 5.2 (the single production site vs. multiple sites within one enterprise being the main distinction) the same actors can be identified at the highest level (see Sect. 5.2.1). As opposed to the refinery manager in the previous case, the problem owner in this case study, the supply chain manager, is responsible not just for overseeing one specific link in the supply chain, but for the operation of a number of sites within one company.

5.3.2 Step 2: System Identification and Decomposition

The *multi-plant enterprise* that will be modelled in this section is a global lube oil supply chain with a global sales department which directly interacts with customers and three production plants at different geographical locations (Zhang et al. 2008). Each production plant itself has several functional departments, each having a specific role and performing certain tasks; the performance of each production plant is a result of its departments' behaviours and their interactions. All production plants operate on a "make-to-order" basis and can produce three types of products after receiving an order from the customer. Each product is produced from eight different raw materials. The goal is to fulfil a set of customer orders in the fastest possible time through assigning them to different production plants and coordinating the behaviour of different departments in each plant.

So, in this enterprise, the involved actors can be viewed on three levels:

Global level — There are three actors at the global level (the customer, the enterprise and the supplier).

Enterprise level — The manufacturing enterprise consists of the global sales department and a number of plants.

Plant level — Each plant has six different functional departments (scheduling, operations, storage, packaging, procurement and logistics).

The behaviour of each of these actors is described in more detail in Fig. 5.2. Based on this system description, an agent-based model with ten agent types has been developed. Table 5.2 presents the social and physical sub-systems (i.e. agents and technologies) in the agent-based model. It is worth mentioning that the manufacturing enterprise is not considered to be a separate agent but a virtual one consisting of the plants' agents and the Global Sales Department (GSD).

5.4 Step 3: Concept Formalisation

The formalised concepts of the socio-technical systems domain, as described in the ontology in Sect. 4.2, were used for both supply chain case studies. For the development of the oil refinery supply chain model no major changes to the generic ontology

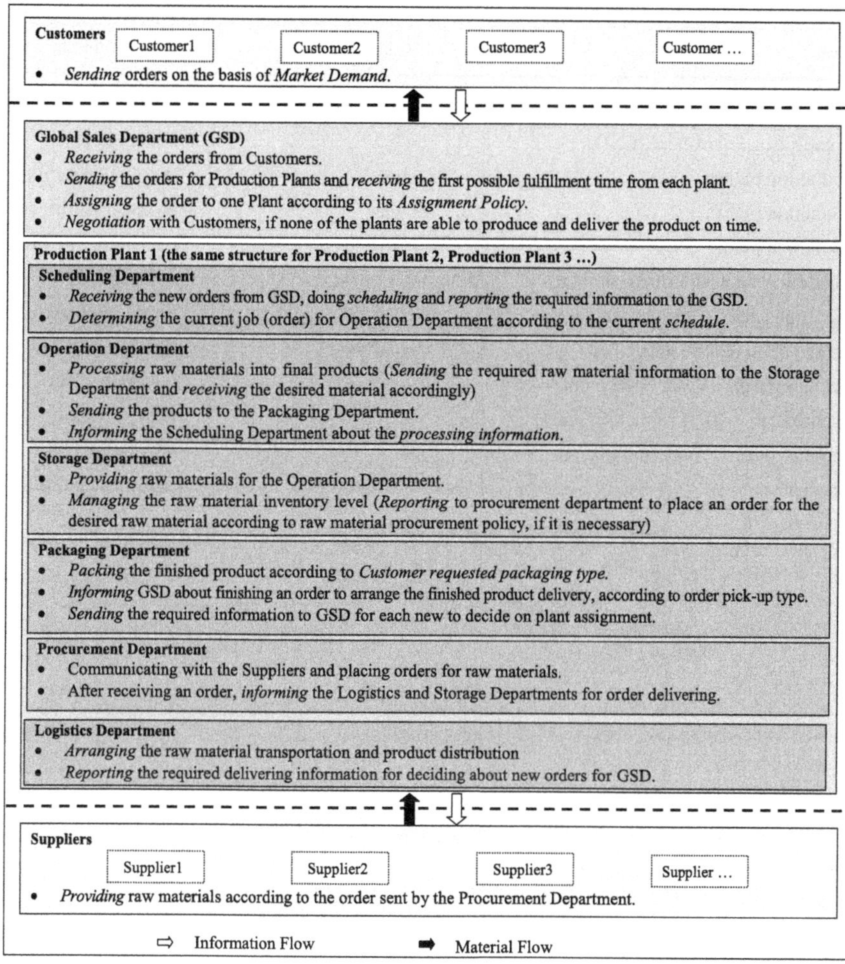

Fig. 5.2 Main actors and their behaviours/interactions in a lube oil multi-plant enterprise

were needed. All the key classes needed to define the system were already in place, based on other case studies (several of which are also included in this book). Mostly minor additions were needed, such as adding properties to the Transport Contract for more detailed registration of transport delays and payment. One of the more significant changes was the adoption of the Ownership class. Initially the Ownership edge represented a link between an Agent and a Physical Node (e.g. the ownership of a Technology by an Agent) only, but from the system decomposition as described in Sect. 5.2.2 it is clear that one Agent can also own another Agent. This is the case when a department of a company is a part of a larger enterprise. The 'to' property was thus relaxed to allow instances of not only Physical Nodes, but also Social Nodes. See Fig. 5.3 for a graphical representation of the new relationship.

Table 5.2 Agents and Physical Nodes and their relationships for the multi-plant enterprise case

Agent	Relationship	Physical Node
Customer	owns/controls	Customer facilities
Global sales department (GSD)	–	not applicable
Production plant	owns	Production facilities
Production plant	owns	Storage facilities
Production plant	owns	Packaging facilities
Scheduling dept. (Production plant)	–	not applicable
Operations dept. (Production plant)	controls	Production facilities
Storage dept. (Production plant)	controls	Storage facilities
Packaging dept. (Production plant)	–	not applicable
Procurement dept. (Production plant)	–	not applicable
Logistics dept. (Production plant)	–	not applicable
Supplier	owns/controls	Supplier facilities

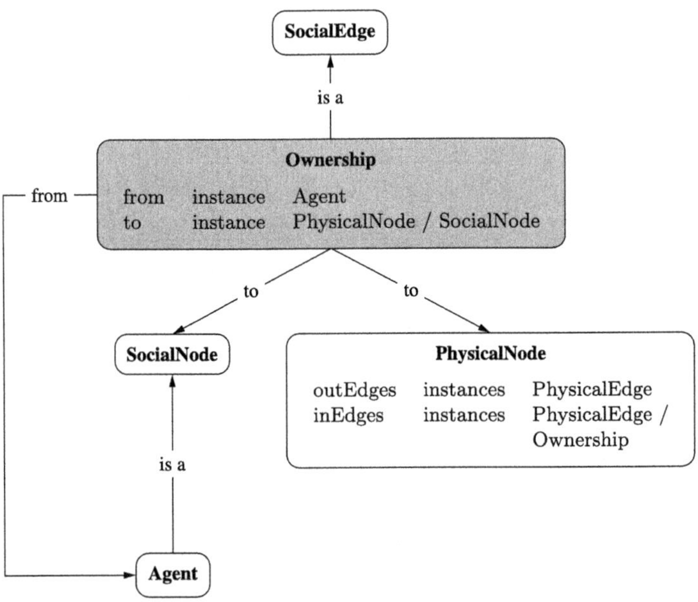

Fig. 5.3 An Ownership edge can connect a Social Node and a Physical Node, forming the link between social and physical networks. In this way we can define an Agent as being the owner of a Technology or another Physical Node. An Ownership edge can also be used between an Agent and another Social Node (e.g. to define that a company consists of several departments or that a department consists of specific people)

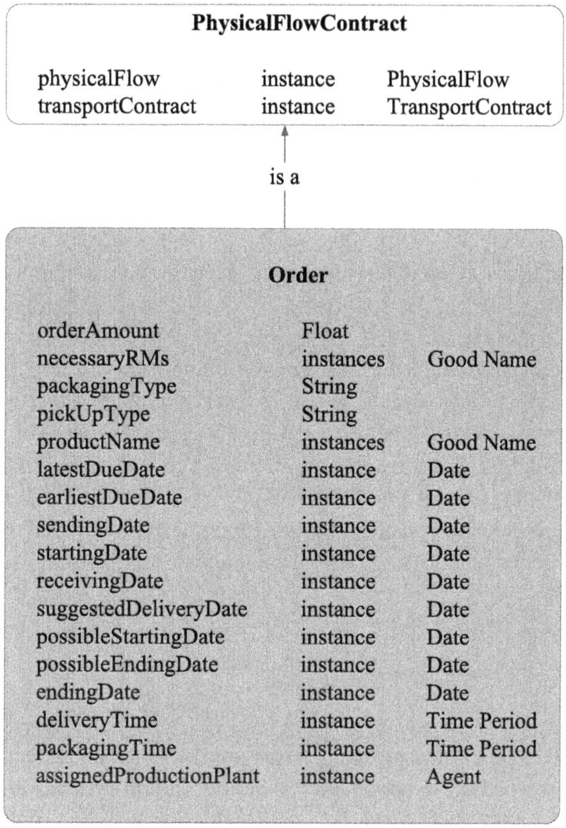

Fig. 5.4 Order ontology

The multi-site case study uses many of the same concepts as the refinery case. Furthermore, the formalisation of several concepts which had not previously been used in other models was required. As one example of a new concept that had to be added, let us take the concept of an *order*. The *order* concept is used to formalise the short-term purchasing transactions between different actors in the supply chain. So, for instance, a customer places an order for a specific final product with a production plant, which in turn further places orders for raw materials with its own suppliers. This concept is especially important when we aim to model a make-to-order production environment (such as a *multi-plant enterprise*). In this type of situation, every good or service can be made only after receiving an order from a customer. In our formalisation we consider *order* to be a sub-class of *PhysicalFlowContract*; therefore, it will inherit the attributes of its super-classes such as *from, to, physicalFlow*. In addition, it has its own specific attributes, including *deliveryTime, latestDueDate, necessaryRMs*[1]. Other attributes of this concept are shown in Fig. 5.4.

[1]RMs stands for Raw Materials.

These new contributions could be re-used in future model developments, if applicable. This could be in additional supply chain models, but these concepts might also be useful for modelling other types of socio-technical systems.

5.5 Step 4: Model Formalisation

In this section the behaviour of the actors in the supply chain is formalised to create the model of the supply chain systems. Following the agent paradigm, the tasks are distributed between the agents. Some tasks therefore have to be split into several subtasks (requiring communication between the agents). A schedule is made so that some processes (e.g. procurement) only occur at certain intervals while others (e.g. production) happen at each time step of the simulation. Events such as the arrival of a VLCC at the jetty are monitored at each time step. After the brief introduction of the roles of the different actors (see for example Sect. 5.2.2), two typical tasks will now be considered and addressed in more detail, namely procurement from the refinery model and order assignment from the multi-site case.

5.5.1 Procurement

The procurement agent is responsible for buying a mixture of crudes well ahead of when these crudes would be required in production. The following narrative describes the steps this actor takes to complete the procurement task.

Before the procurement agent can place an order from the suppliers, it needs to determine which crudes to buy and how much of each product. First, the procurement department asks the sales department what the predicted demand is for each of the products produced by the refinery (i.e. gasoline, kerosene, diesel and fuel oil). After it has received the volumes that are expected to be produced, the procurement department determines what product has the highest crude demand, as this will be crucial in choosing the crude mixture and amounts. The average yields of the possible crude mixtures are used to calculate which product will require the highest amount of crude oil. The operations department is asked which recipe it will use to produce the product with the highest crude demand, considering all other products bought by the consumers as by-products, because this approach ensures that at least enough is produced. Next, from the recipe selected by the operations department, it can be deduced which crudes need to be bought and in which quantities.

The procurement department has multiple strategies to decide how much crude to buy, two of which are included in this model. In the first policy the amount of each crude to buy is solely determined by the recipe selected and the total amount needed to produce the expected order. In the second, more advanced, policy, the procurement department takes into account the current levels in the storage tanks (obtained from the storage department) as well as the amount in the tank which is already reserved for the current production cycle (information obtained from the operations department). A safety stock level is kept as a minimum volume to be kept in the storage tank. After it has been determined which crudes need to be bought and how much of each one of these crudes is required, the procurement department approaches the market to find the best seller. It does this by asking all potential sellers for a contract and selecting the cheapest one, which is signed to form a trade agreement.

There are a number of prerequisites, such as that the sales department has completed its forecast of the demand. For the process to be completed several other tasks need to be performed, including asking the logistics department to arrange transport for the crudes that have been bought. These tasks are scheduled to be performed by other agents.

This narrative can be translated to pseudo-code as shown in Algorithm 5.1, considering only the first procurement strategy and omitting the selection of the best contract for simplification.

All other behaviours in the model are also split up in a similar fashion between the different agents. Another example of this is the selection of the production mode by the operations department, which is based on the forecasts made by the sales department and the crudes selected by the procurement department. Where possible, modelling the behaviour of the agents was based on existing behavioural rules for trading. However, additional rules had to be implemented: for example, for various procurement policies (e.g. forecasting of demand deciding on procurement), for scheduling which Operational Configuration to use, and for the activities of the jetty which had not been used in earlier models (van Dam et al. 2009a).

5.5.2 *Order Assignment*

Similar to those in the *oil refinery case*, the actors in a multi-plant enterprise do different sorts of activities and participate in different processes. Basically, the main determinant of their behaviour is the policies they have. As an example, for a procurement agent the "reorder point" is a policy for managing the raw material level and, according to it, this agent must contact another agent (supplier) to place raw material orders at appropriate times.

To illustrate how different agents may participate in a specific process and coordinate their activities in the multi-plant enterprise, the "order assignment process" is described here:

After receiving a new order from customers, the GSD will assign it to one of the available production plants. For this purpose, the GSD sends the order information to the scheduling department of production plants. Each scheduling department then replies with the earliest date by which the plant can make the product and deliver it to the customer. Based on the replies, the global sales department will assign the customer order to one production plant according to its assignment policy. This assignment policy can be a time-related or cost-related policy. For example, if the

Algorithm 5.1 Procurement process

```
get forecastDemands from SalesDept

set productWithHighestCrudeDemand to empty
set highestCrudeDemand to 0

for goodName from forecastDemands
do
    set yield to averageYields for this goodName
    set demand to forecastDemands for this goodName

    if (demand/yield) > highestCrudeDemand
        set highestCrudeDemand to demand/yield
        set productWithHighestCrudeDemand to goodName
    end if
end for

get operationalConfiguration from
    operationsDepartment

if procurementStrategy equals 1
    for crude from operationalConfiguration
    do
        set amountToBuy to operationalConfiguration.
            relativeAmount * highestCrudeDemand

        buy amountToBuy of crude

    end for
end if
```

assignment policy is "First Completion Date Policy", then the order is assigned to the production plants with first possible fulfilment date. If none of the plants are able to produce and deliver the product on time, the GSD will start the negotiation with customer for extending the order due date.

This process can be described with Algorithm 5.2. This algorithm describes the sequence of activities in the "order assignment process" and the actors which perform these activities. The behaviour of different actors and other processes in the system (such as "Order Replenishment Process" or "Abnormal Situation Process") can be expressed in similar algorithms and pseudo-code.

Algorithm 5.2 Order assignment process

```
announceOrder by globalSalesDepartment

for ProductionPlant from AllAvailableProductionPlants
do
        obtain Scheduling Department of Production Plant
  obtain Schedule from Scheduling Department for non-Assigned Orders and new Orders
  obtain Processing Timing from Scheduling Department for new Orders
end for

determineFirstPossibleDate for newOrder by globalSalesDepartment

if LatestDueDate of newOrder >= firstPossibleDate
        assignTheOrderToTheFPTPlant
else
        startNegotiationOnExtendingDueDate
end if
```

5.6 Step 5: Software Implementation

Both models have been implemented in a similar fashion using Java. The software implementation is based on the software framework including shared basic classes for the ontology concepts and several aids for visualisation and data analysis. This shared framework was updated following the specifications for these particular case studies. For example, following the extension of the Ownership class in Sect. 5.4, the Java class representing an agent was adjusted to implement the consequences of an agent being owner by another agent. For example, if an agent which is a subsidiary of another agent has to make a payment, it will use the "wallet" of the agent which owns it (this may be recursive).

There are no specific hardware requirements, as the models are designed to run on a standard desktop computer.

5.7 Step 6: Model Verification

The oil refinery model has been tested extensively. A benchmarking study was executed in which the agent-based model was compared with an equation-based model

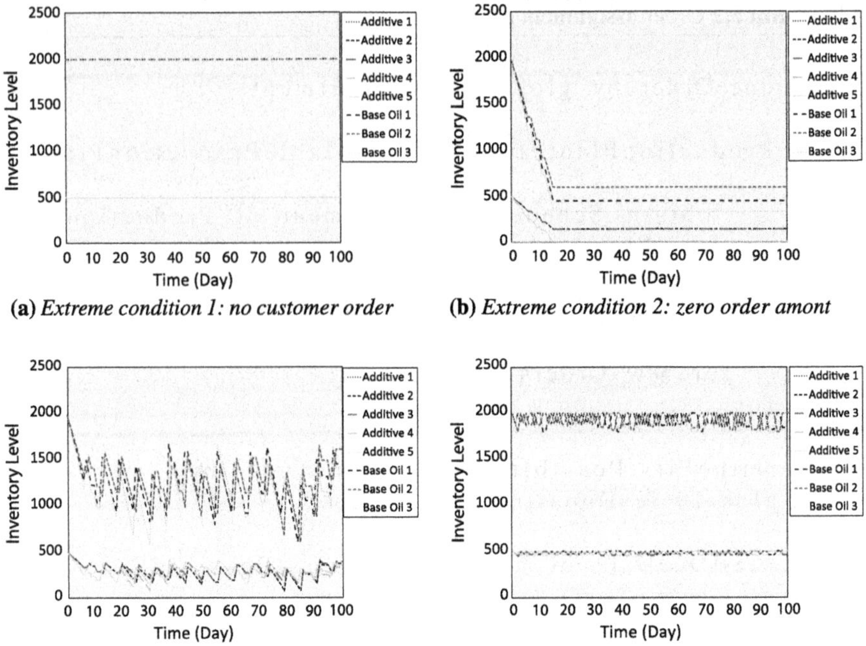

(a) *Extreme condition 1: no customer order*

(b) *Extreme condition 2: zero order amont*

(c) *Extreme condition 3: high order rate*

(d) *Extreme condition 4: high order and immediate delivery*

Fig. 5.5 Model behaviour (inventory level for production plant 1) under different extreme condition tests

of the same supply chain (van Dam et al. 2009a). The aim of the benchmarking study was to compare the modelling paradigms (i.e. the agent-based and equation-based paradigms) and to learn about the advantages and disadvantages of the different approaches. In order to benchmark the modelling paradigms, it was necessary to demonstrate that the models under study are comparable. During this process a number of differences between the results of earlier versions of the model were identified, which could then be checked in more detail. Several mistakes made during the formalisation and implementation phases were then able to be corrected in this way. Finally, a numerical analysis was performed, proving that the two models show the same behaviour (van Dam et al. 2009a), raising confidence that the agent-based model was implemented correctly.

For the multi-plant enterprise case, tests of extreme conditions and behaviour sensitivity were conducted. In these tests—in which selected parameters had extreme input values, such as zero—the model should behave according to our expectations. Figure 5.5 shows the inventory levels of different raw materials for production Plant 1 under four extreme conditions, as discussed below:

- The test related to Fig. 5.5a was conducted with the assumption of no consumer orders being placed for the duration of the simulation. So, as we expected, the inventory level for raw materials is unaltered. The other two production plants produced similar results.

- In Fig. 5.5b, we set the reorder amount to zero. The results show that the production plant will use its initial raw material in the storage facilities, but as there is no replenishment, the raw material level will be zero for the rest of simulation time.
- Figure 5.5c shows a case in which the reorder point is set to 0.9; i.e. when the raw material inventory level falls below 90 percent of storage capacity, a new raw material order is placed. The expected result is frequent raw material orders, and this is confirmed in the figure.
- Figure 5.5d shows the results of a test similar to the previous case but with immediate raw material delivery (the raw material delivery time is set to zero). Again, the model behaves as expected and the stock is always near maximum capacity.

Similar tests can be done by using extreme values for other parameters (e.g. order processing time, storage capacity or plant availability) to study whether those factors have the expected effect on the model outcomes.

5.8 Step 7: Experimentation

Now that the models have been designed and implemented, they can be used to run experiments that can help answer the main questions posed in Sects. 5.2.1 and 5.3.1. Below, we first address the design of the experiments for the oil refinery case study, followed by a discussion for the multi-site case.

5.8.1 Experimental Setup for the Oil Refinery Supply Chain

In a similar way to that of the problem formulated in Sect. 5.2.1, an experiment will now be now set up to evaluate the best response to a *disturbance*, namely a delay in a shipment of raw materials.[2] There are different options for the criteria with which to choose the best alternative. As examples, one can look at the overall profit of the refinery (for a certain time frame), profit during the production cycle affected by the disturbance, other financial measures, but also non-economical criteria such as customer satisfaction. Profit P of the refinery was chosen, at 14 days after a disruption took place. This means that the effect of a disturbance will be simulated over the next two cycles of operation, during which new raw materials are ordered and products are dispatched. It is assumed that the impact of the disturbance will have worn off by then.

The objective function for a period of 50 days is defined as follows:

[2]For simplicity, it is assumed that the magnitude of the disturbance is known as soon as the disturbance occurs. In reality, this may involve uncertainty. Furthermore, currently delays are in the order of magnitude of days, but the granularity could be adjusted so that a delay could be expressed in parts of a day (e.g. hours) instead of full days of 24 hours.

$$\begin{aligned}\max_{\bar{x}} P(\bar{x}, \bar{d}) = \sum_{t=1}^{50} \big(& Income^{t}_{\text{sales}}(\bar{x}, \bar{d}) - Cost^{t}_{\text{procurement}}(\bar{x}, \bar{d}) \\ & - Cost^{t}_{\text{transport}}(\bar{x}, \bar{d}) - Cost^{t}_{\text{maintenance}}(\bar{x}, \bar{d})\big) \\ & + Value^{t=50}_{\text{product stock}}(\bar{x}, \bar{d}) + Value^{t=50}_{\text{raw materials}}(\bar{x}, \bar{d}) \end{aligned} \tag{5.1}$$

The function for the transportation costs, and therefore profit, is discontinuous because these are a function of the amount of crude procured, the capacity of the ships and the travel time. The transport cost is calculated per vessel, which could either be a very large crude carrier for long-distance shipping or a general purpose tanker with much smaller capacity for short hauls in the case of emergency procurement, and the unit landed-cost of crude (i.e. procurement plus transport costs) therefore follows a saw-tooth pattern. This discontinuity makes it more difficult to determine the right amount to buy, especially in combination with other measures such as switching to another recipe in the refinery.

In Eq. (5.1) the monetary value of the product inventories and raw material stocks at the end of the simulation run are included. The consequences of the disruption to future cycles within the time horizon are thus included in the cost function. One such example might be that if the response is to switch to another mode of operation without any emergency procurement, it is possible that during a later cycle the planned operation cannot be met. However, no new decisions following such new disturbances are assumed; a single response is formulated.

Faced with a disturbance, the problem owner must make a number of choices. Firstly, it has to determine if the disturbance has a significant effect on the operation of the supply chain. If the effect is deemed minor, no action may be necessary, but if it is not able to execute the previously planned schedules due to insufficient crude, corrective action may be required. A disturbance in the supply of crudes can be addressed by changing the operating mode or the throughput, or by emergency crude procurement. Often a combination of these actions is needed.

For the emergency procurement *EmPr*, the procurement department can contact a local supplier to buy crude at a much higher price but with a shorter lead time. The procurement department has to ask the logistics department for the expected delay to be able to make this decision. Furthermore, the operations department can choose to change the operational configuration (*COC*), meaning that a different recipe is selected using crudes that are still in stock but which result in yields that are not ideal compared to the scheduled operation. Finally, the operations department can change the operational scale (*COS*) to run the refinery at a lower throughput, producing fewer end products but avoiding having to shut down the plant when crude runs out (or postponing plant shutdown, for example to allow emergency procurement crudes to arrive).

The degrees of freedom are thus defined as follows:

$$\bar{x} = (EmPr_1, EmPr_2, EmPr_3, EmPr_4, EmPr_5, COC, COS) \tag{5.2}$$

For emergency procurement the degree of freedom for each of the five crudes is between 0 kbbl and the amount that could reasonably be available on short notice, which is assumed to be 600 kbbl.[3] Furthermore, the number of different recipes in the refinery is assumed to be four, one of which is always selected as the current operational configuration. The CDU in the refinery has a minimum capacity as one of its design parameters; below 40 % of the maximum throughput the process no longer works and the refinery has to be shut down. These constraints are defined as follows:

$$0 \leq EmPr_i \leq 600 \quad \text{in kbbl, for each of the } i = 5 \text{ crudes} \tag{5.3}$$

$$COC \in \{R1, R2, R3, R4\} \quad \text{discrete choice between operating modes} \tag{5.4}$$

$$40 \leq COS \leq 100 \quad \text{percentage of CDU throughput capacity} \tag{5.5}$$

The scope is limited to disturbances $\bar{d}$ dealing with the supply of crude oil to the crude distillation units. These are defined by:

$$\bar{d} = (ShipDelay, StorageProblem) \tag{5.6}$$

with:

$$ShipDelay \in \{0, 1, 2, \ldots, n\} \quad \text{in days, for 1 ship for 1 cycle} \tag{5.7}$$

$$StorageProblem \in \{0, 1\}^m \quad \text{for each of the } m = 5 \text{ crude storage tanks} \tag{5.8}$$

For this case study it is assumed that a disturbance to the system occurs on day $t = 22$. A ship at sea is delayed for 30 days, but there are no problems with any of the storage units. This means that the disturbance is defined as $\bar{d}(22) = (30, 0, 0, 0, 0, 0)$.

The Nelder-Mead optimisation method (Nelder and Mead 1965) is used here as the search strategy, to determine *which* experiments need to be performed with the model. The method is based on identification of the best, the worst, and the second worst outcomes in each iteration for the pre-defined simplex (a set of experiments). See the box below for more details. This method has not only been chosen because of its ability to deal with discontinuous objective functions, but also to illustrate how an optimisation method that is commonly used in process systems engineering using mathematical models (Biegler and Grossmann 2004) or samples from experiments in a real system can also be a powerful approach when combined with an agent-based model.

[3]kbbl stands for 1000 standard oil barrels.

Nelder-Mead optimisation method

In the Nelder-Mead optimisation method (Nelder and Mead 1965), an initial simplex S is defined as a convex hull with $n + 1$ vertices $\{\bar{x}_j\}_{j=1}^{n+1}$ in an n-dimensional space $\mathbb{R}^n$ (with n equal to the number of degrees of freedom in $\bar{x}$). These vertices satisfy the non-degeneracy condition, meaning that the volume of the simplex hull is non-zero. For every next iteration $j + 1$, the values for $\{\bar{x}_j\}_{j+1}^{n+1}$ are determined by comparing the objective-function values followed by replacement of the worst vertex by another point. The simplex adapts itself to the local landscape and finally contracts to the (local) optimum. In other words, for an initial collection of chosen values for each of the variables one may wish to adjust (i.e. the "degrees of freedom") it is determined, in this case through simulation, what the performance of the system is. The search algorithm then determines, based on these findings, which variables to adjust and in which direction, through extrapolation of the previous generation. By repeating this process and running new simulations using the values suggested by the algorithm, the variable landscape is explored until a suitable solution is found.

An initial simplex S for a 6-dimensional space (the number of degrees of freedom in $\bar{x}$) for the Nelder-Mead optimisation method needs to contain 7 vertices. Table 5.3 (rows 1 to 7) on page 175 shows the initial values that were chosen, based on a first analysis of the problem by an expert. These values for the degrees of freedom provide enough variation for the search algorithm to proceed. The stop condition for the search is when all $\bar{x}$ in the population have prevented a shutdown of the refinery and no better values for the criteria are found through a new iteration.

5.8.2 Experimental Setup for the Multi-plant Enterprise

The agent-based model developed for the multi-plant enterprise can be used to set up many experiments and to study important factors that influence the behaviour of each plant separately as well as the performance of the enterprise as a whole. For this purpose we need to define some performance indicators at the enterprise and plant levels. In general, the performance of a supply chain can be analysed in terms of customer service (e.g. tardiness, number of late orders), financial aspects (e.g. profit, overall operational cost), or a combination of both. The performance indicators considered in the experimental setup in this chapter are *number of late orders*, *total tardiness (in days)* and *average inventory level*. It should be stressed that the flexibility of an agent-based model guarantees an easy extension of the original performance indicator if it appears that additional analysis is necessary. The experiments with the developed model can be defined by implementing different behaviours for the different departments (i.e. changing their properties or their working policies).

Meanwhile, the objective of the experimental setup can be related to the normal operation of this multi-plant supply chain (i.e. to evaluate the dynamic behaviour of supply networks under different conditions or to study the effect of different policies on system performance in order to find the most appropriate policy). For instance, the effect of changing the reorder point (procurement policy) on system performance is studied. For this purpose, this parameter was changed from 10 to 25 percent, so as to find an appropriate value that minimises the *number of late orders* and *total tardiness*. Similarly, changing the policies for different departments may have many different effects on different system performance indicators, and accordingly, the normal operation of the supply chain under different scenarios can be analysed.

In addition to those for the normal behaviour of a multi-plant enterprise, many experiments can be formulated to study the enterprise performance during different, abnormal, situations and to find effective strategies to cope with them. As an example, we assumed an unexpected shutdown in Production Plant 1 at the 70th day of the time horizon. Consequently, starting with the 71st day, all orders must be fulfilled by Production Plants 2 and 3. As expected, the overall performance was affected by this disruption, resulting in a number of non-finished and late customer orders, and the nominal optimum reorder point was not optimal for the disturbed situation. To handle this disruption, the enterprise could adapt its different policies (e.g. its procurement policy) by changing its nominal reorder point. The effect of this adaptation is also studied with the model.

5.9 Step 8: Data Analysis

Next, the results of the experiments described in Step 7 are presented and analysed for both case studies.

5.9.1 Delay in Shipment in the Oil Refinery Supply Chain

Figure 5.6a illustrates the crude stocks of the refinery under normal operation, and Fig. 5.6b shows the operational scale (both planned and actual throughput) over time after disturbance $\bar{d} = (30, 0, 0, 0, 0, 0)$ which occurs on day $t = 22$. This disturbance resulted in a loss of \$24 million because of lost production hours, as illustrated by the gap between planned and actual throughput. Furthermore, the agent-based simulation model supports the decision about which response is the most appropriate given the many degrees of freedom. The change in operational configuration was not included in this experiment; rather, the planned production mode (in this case $R4$) was used. For each possible $COC \in \{R1, R2, R3, R4\}$ an optimisation for $EmPr_i$ and COS had to be performed because choosing a different recipe would necessarily influence the criterion surface. Preliminary results for the decision on emergency

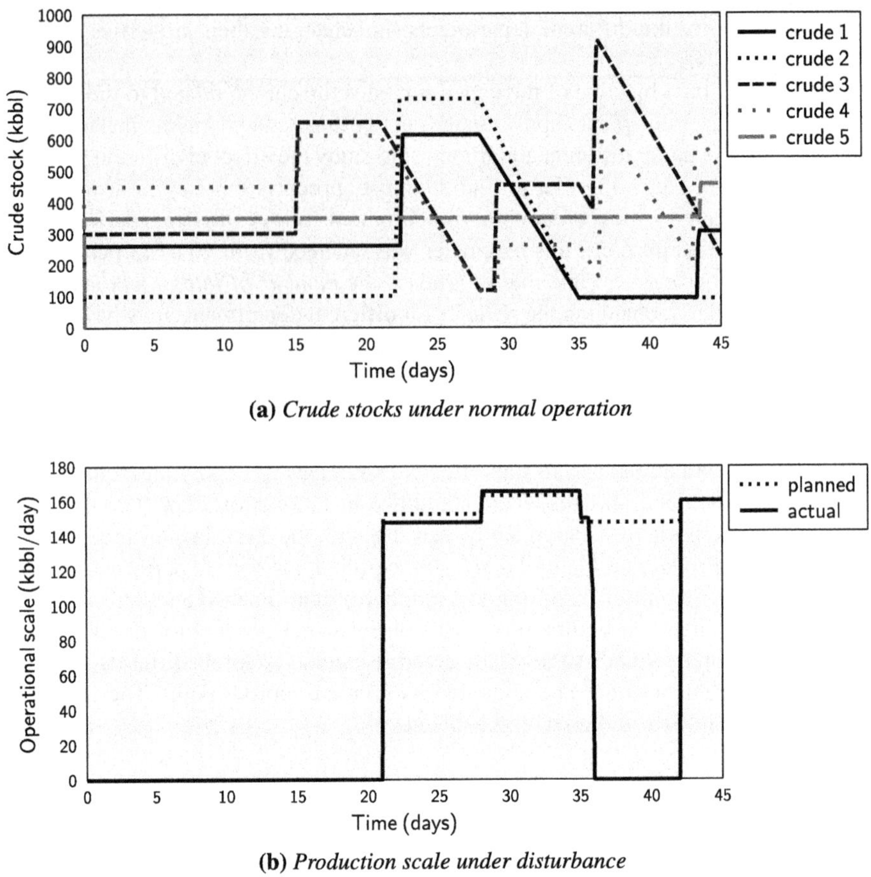

Fig. 5.6 Results from the oil refinery supply chain case study

procurement and the change of operational scale are shown in Table 5.3. After 20 iterations no further improvement was made, so the search was terminated. The proposed solution prevents a shutdown of the refinery by purchasing emergency crudes to make up for the delayed shipment and by slightly reducing the throughput. The loss caused by the disruption was reduced by \$14.7 million (excluding penalties to be paid by the shipper for delays) which is more than 60% of the \$24 million in damage caused by the disturbance (van Dam et al. 2009b).

5.9.2 Normal and Abnormal Behaviour Analysis for the Multi-plant Enterprise

Figure 5.7 shows the effect of changing the reorder point (procurement policy) on plant-level and enterprise-level performance. Accordingly, as the reorder value in-

Table 5.3 Outcome of the Nelder-Mead simplex optimisation algorithm with the agent-based model

	$EmPr_1$	$EmPr_2$	$EmPr_3$	$EmPr_4$	$EmPr_5$	COS	$P(\bar{x}, \bar{d})$
1	0	0	0	0	0	60	−5.69 E8
2	0	0	100	100	0	60	−3.65 E8
3	0	0	200	300	0	60	−1.61 E8
4	0	0	300	250	0	55	−1.64 E8
5	0	0	300	250	0	50	−6.19 E7
6	100	100	100	100	100	50	−3.70 E8
7	10	10	500	600	10	50	1.39 E8
final	4	4	349	382	4	54	1.45 E8

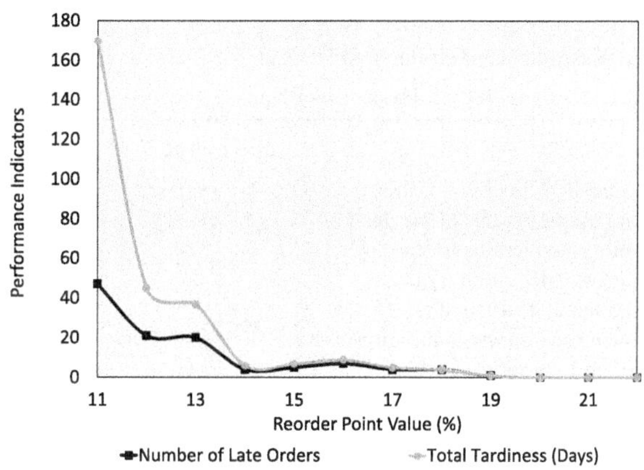

Fig. 5.7 The effect of the procurement policy (reorder point) on overall system performance (total tardiness and number of late orders)

creases, the overall performance of the enterprise improves (*total tardiness* and *number of late orders* decrease); since without raw material, the operation department must pause its execution of an order and wait to receive the raw materials from the supplier. This causes delay in fulfilling customer orders that are assigned to that production plant. Generally speaking, with a higher reorder point the raw material availability can be a less important bottleneck for the production plant. Figure 5.7 also suggests considering 20 percent as an optimum reorder point for this set of customer orders. This is because 20 percent is the first reorder value at which both *total tardiness* and *number of late orders* are zero; all customer orders can thus be fulfilled without any delay.

Table 5.4 illustrates how an unexpected shutdown in Production Plant 1 at the 70th day of the time horizon will affect the performance of multi-plant enterprise. The assumption was that all production plants were working according to their optimum reorder point (20 percent). Because of this plant disruption, there are 14 late orders with 34 total tardy days. As mentioned in the previous section, one way of managing this abnormality is to adapt the policies for different departments. One possibility would be to change the procurement policy for Plants 2 and 3 immedi-

Table 5.4 Performance data for the enterprise in a normal and an abnormal situation (Plant 1 shutdown); the reorder point in both cases is 20 percent

Performance indicator	Normal operation	Abnormal situation
Orders assigned to ProductionPlant1	33	22
Orders assigned to ProductionPlant2	33	39
Orders assigned to ProductionPlant3	34	39
Orders assigned to All Production Plants	100	100
Late Orders by ProductionPlant1	0	0
Late Orders by ProductionPlant2	0	7
Late Orders by ProductionPlant3	0	7
Late Orders by All Production Plants	0	14
Total tardiness for ProductionPlant1	0	0
Total tardiness for ProductionPlant2	0	16
Total tardiness for ProductionPlant3	0	18
Total tardiness for All Production Plants	0	34

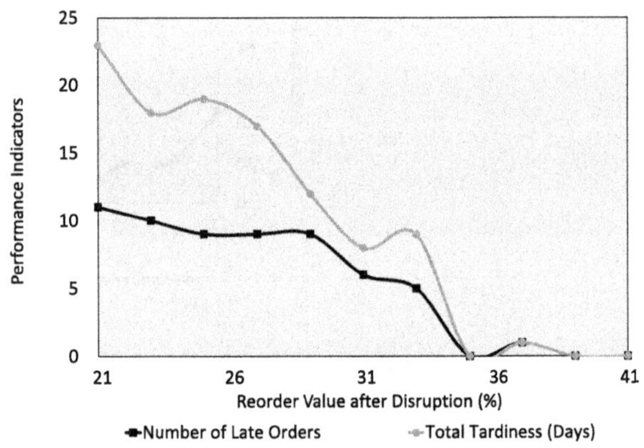

Fig. 5.8 The effect of the procurement policy (reorder point) on overall system performance (total tardiness and number of late orders) during an abnormal situation (Plant 1 shutdown)

ately following the disruption in Plant 1. The relation between changing the reorder point for Production Plants 2 and 3 after disruption and the total tardiness as well as the number of late orders is shown in Fig. 5.8. In general, increasing the reorder point improves these two performance indicators; but on the other hand, a higher reorder point will increase *raw material holding cost*, too (Fig. 5.9). In total, if the reorder point is increased, the *average inventory level* must also be increased; but the inventory level is also dependent on the number of completed orders. Having more finished orders means more raw material consumption and a lower average inventory level. Based on this analysis, setting the reorder point for the two available production plants to a value of approximately 35 percent will result in desirable levels for both *customer satisfaction* and *raw material holding cost*. Of course, for a real case, according to the storage-related costs, a more precise value can be determined. Meanwhile, an enterprise can react to an abnormal situation with many

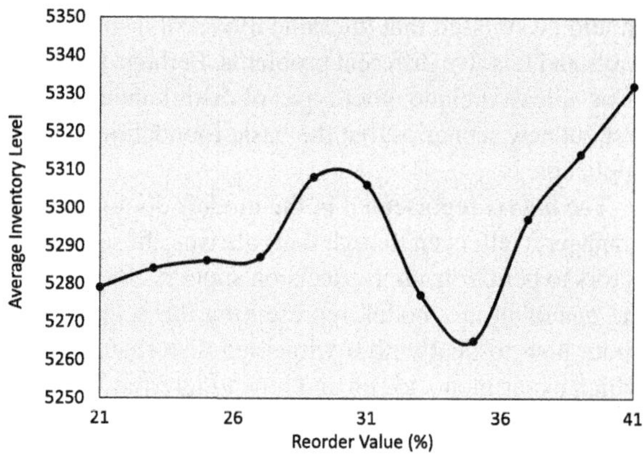

Fig. 5.9 The effect of the procurement policy (reorder point) on overall system performance (average inventory level) during an abnormal situation (Plant 1 shutdown)

other strategies, and changing the reorder point can be a part of a comprehensive crisis management package.

5.10 Step 9: Model Validation

Following the views on validation presented in Sect. 3.10 in Part I, the main concern is how to be certain that the model is applicable for its domain and can aid the understanding of the problem under study. This application relevance can be shown by different experiments done with the model and the insights gained regarding, for example, the dynamic behaviour of the system. The previous subsections in this chapter show the application of both developed models for the problems defined in Sects. 5.2.1 and 5.3.1 by defining different experimental setups. Furthermore, when it comes to expert validation, for both models presented in this chapter, the domain expert validation was done in all stages of model development. As mentioned in Sect. 5.7, the refinery supply chain model was benchmarked against another model. That equation-based model has been validated against the real system and was applied in order to offer decision support (Pitty et al. 2008). Since the agent-based model of the refinery supply chain was successfully compared with an already validated mathematical model, the agent-based model that was based on it can be considered valid also. Similarly, feedback from domain experts on the multi-plant case gives confidence that the model is valid and useful.

5.11 Step 10: Model Use

The models presented in this chapter have been used for *offline* decision support, meaning that they are used to test different scenarios in a simulation, after which the "best" solution can be implemented in the real system. In an *online* controller the results of the model could be directly used to affect the real system. However, it should be stressed that the same model of a supply chain could be used in different ways and to solve different problems. Perhaps some changes would have to be made to be able to include other types of disturbances and add new degrees of freedom to test out new scenarios, but the basic foundation of the model will remain useful and applicable.

The *actors* represented in the models do not include the role of the supply chain manager itself, even though that role was chosen as the problem owner and the main actors to benefit from the decision support offered by these models. This means that the *agents* in the model, representing the actors, do not themselves take decisions about how to deal with a witnessed disturbance, but it is up to the *user* to decide which experiments to run and how to interpret the outcomes.

Various roles of simulation in supply chain management can be considered, including (Merkuryev et al. 2009):

- Back-up decision-making
- Validate algorithms
- Convince users of supply chain management approaches
- Educate

Models of supply chains have been successful in offering decision support. As an example, consider early supply chain models which could be used to visualise the so called *bull whip effect* in which the effects of decisions are propagated through the supply chain (Forrester 1958). Furthermore, different policies and algorithms have been tested with such models (Julka et al. 2002b). Simulations of supply chains have also been use for educational purposes to elucidate the dynamics within a chain of different entities (Holweg and Bicheno 2002).

One specific issue with the visualisation of data and simulation outcomes of supply chain models is making sure that justice is done to the wide range of actors or geographical locations. As stated above, supply chains are complex systems with multiple social entities intertwined with many physical systems. Solving a problem in one aspect of the supply chain (e.g. a red alert for a storage tank) might require actions elsewhere in the supply chain. As of yet this is an unsolved challenge in the two models presented in this chapter. This would, however, be an important step towards the development of decision support tools which can be used by different stakeholders, even if they are not modellers themselves.

5.12 Conclusions

This chapter has described the application of agent-based modelling as a promising approach to capturing the dynamic and complex behaviour of an industrial supply chain. Meanwhile, the two cases studies presented in this chapter have shown that models developed using the modelling steps from Chap. 3 can be used to support decision-makers for managing their supply chains, both during normal operation and when facing abnormalities.

Each supply chain is a decentralised, complex and adaptive system with many heterogeneous actors and physical components interacting through different flows, including material, information, monetary and social flows. Furthermore, decision-making in a supply chain is distributed among different actors, and each of these actors has its own objectives and procedure for decision-making. The collective decisions made by these autonomous actors at various levels of the system result in the overall system behaviour. So the need for a modelling approach that can capture all these complexities and this decision structure is clear. As discussed in this chapter, agent-based modelling seems to be one of the best candidates for addressing these issues. In addition, agent-based modelling is flexible and can define a broad range of experiments with different scenarios to answer "what if" questions; this is critical for decision support under disruptions or in the design phase. The models used in this chapter were developed in a bottom-up fashion, making it relatively simple to change the configuration: it is easy to include new actors in the system (e.g. more suppliers with different prices and lead times in the supply chain) or to adjust the physical configuration (e.g. extra storage tanks for the refinery or production plants).

However, there are some barriers that may affect the willingness to use this modelling approach: one of the main challenges concerns validation, and this is one of the main sources for much of the criticism that agent-based modelling has received as a research method in the literature. Obviously, these criticisms are not solely about using agent-based modelling for supply chain management. In addition, considering the novelty of this approach (especially as a decision support methods for complex adaptive systems), there is a need for more research on finding some universally accepted validation methods.

Acknowledgements The authors would like to express their thanks to Rajagopalan Srinivasan and Arief Adhitya for their support in the development of the models used in both case studies.

References

Behdani, B., Lukszo, Z., Adhitya, A., & Srinivasan, R. (2010a). Agent-based modelling to support operations management in a multi-plant enterprise. *International Journal of Innovative Computing, Information and Control*, *6*(7), 1–12.

Behdani, B., Lukszo, Z., Adhitya, A., & Srinivasan, R. (2010b). Performance analysis of a multi-plant speciality chemical manufacturing enterprise using an agent-based model. *Computers & Chemical Engineering*, *34*(5), 793–801.

Bhatnagar, R., Chandra, P., & Goyal, S. K. (1993). Models for multi-plant coordination. *European Journal of Operational Research*, *67*(2), 141–160.

Biegler, L., & Grossmann, I. (2004). Retrospective on optimization. *Journal of Computers and Chemical Engineering*, *28*(8), 1169–1192.

Forrester, J. (1958). Industrial dynamics: a major breakthrough for decision makers. *Harvard Business Review*, *36*(4), 37–66.

Gjerdrum, J., Shah, N., & Papageorgiou, L. G. (2000). A combined optimisation and agent-based approach for supply chain modelling and performance assessment. *Production Planning & Control*, *12*, 81–88.

Holweg, M., & Bicheno, J. (2002). Supply chain simulation—a tool for education, enhancement and endeavour. *International Journal of Production Economics*, *78*(2), 163–175.

Julka, N., Srinivasan, R., & Karimi, I. (2002a). Agent-based supply chain management—1: framework. *Computers & Chemical Engineering*, *26*(12), 1755–1769.

Julka, N., Karimi, I., & Srinivasan, R. (2002b). Agent-based supply chain management—2: a refinery application. *Computers & Chemical Engineering*, *26*(12), 1771–1781.

Kaihara, T. (2003). Multi-agent based supply chain modelling with dynamic environment. *International Journal of Production Economics*, *85*(2), 263–269.

Lee-Post, A., & Chung, C. H. (2008). *Systems for supporting operations management decisions*. Berlin: Springer.

Mele, F. D., Guillén, G., Espuña, A., & Puigjaner, L. (2007). An agent-based approach for supply chain retrofitting under uncertainty. *Computers & Chemical Engineering*, *31*(5–6), 722–735.

Merkuryev, Y., Merkuryeva, G., Piera, M., & Guasch, A. (2009). *Simulation-based case studies in logistics*. Berlin: Springer. ISBN: 978-1-84882-186-6.

Min, H., & Zhou, G. (2010). Supply chain modeling: past, present, and future. *Computers and Industrial Engineering*, *43*(1–2), 231–249.

Moyaux, T., Chaib-draa, B., & D'Amours, S. (2006). *Supply chain management and multiagent systems: an overview*. Berlin: Springer.

Nelder, J. A., & Mead, R. (1965). A simplex method for function minimization. *Computer Journal*, *7*(4), 308–313.

Pitty, S. S., Li, W., Adhitya, A., Srinivasan, R., & Karimi, I. (2008). Decision support for integrated refinery supply chains—1. Dynamic simulation. *Computers & Chemical Engineering*, *32*(11), 2767–2786.

Siirola, J. D., Hauan, S., & Westerberg, A. W. (2003). Toward agent-based process systems engineering: proposed framework and application to non-convex optimization. *Computers & Chemical Engineering*, *27*(12), 1801–1811.

Srinivasan, R., Bansal, M., & Karimi, I. (2006). *A multi-agent approach to supply chain management in the chemical industry*. Berlin: Springer.

Swaminathan, J. M., Smith, S. F., & Sadeh, N. M. (1998). Modeling supply chain dynamics: a multiagent approach. *Decision Sciences*, *29*(3), 607–632.

Tayur, S., Ganeshan, R., & Magazine, M. (1999). *Quantitative models for supply chain management*. Norwell: Kluwer Academic.

van Dam, K. H. (2009). *Capturing socio-technical systems with agent-based modelling*. PhD thesis, Delft University of Technology, Delft, the Netherlands. ISBN 978-90-79787-12-8.

van Dam, K. H., Adhitya, A., Srinivasan, R., & Lukszo, Z. (2008). Benchmarking numerical and agent-based models of an oil refinery supply chain. In B. Braunschweig & X. Joulia (Eds.), *Computer-aided chemical engineering: Vol. 25. Proceedings of the European symposium on computer aided process engineering ESCAPE 18* (pp. 623–628). Lyon: Elsevier.

van Dam, K. H., Adhitya, A., Srinivasan, R., & Lukszo, Z. (2009a). Critical evaluation of paradigms for modelling integrated supply chains. *Journal of Computers and Chemical Engineering*, *33*(10), 1711–1726. 2009.01.023. doi:10.1016/j.compchemeng.

van Dam, K. H., Lukszo, Z., & Srinivasan, R. (2009b). Abnormal situation management in a refinery supply chain supported by an agent-based simulation model. In R. M. de Brito Alves, C. A. O. do Nascimento, & E. C. Biscaia Jr. (Eds.), *Proceedings of the 10th international symposium on process systems engineering—PSE2009*, Salvador, Bahia, Brazil.

Zhang, H., Wong, W. K., Adhitya, A., & Srinivasan, R. (2008). Agent-based simulation of a speciality chemicals supply chain. In *Proceedings of the first international conference on infrastructure systems (INFRA 2008): building networks for a brighter future*, Rotterdam, The Netherlands.

Chapter 6
An Agent-Based Model of Consumer Lighting

E.J.L. Chappin and M.R. Afman

Abstract With the aim of better understanding the consequences of the EU ban on incandescent lamps, an agent-based model has been developed in which consumer behaviour in the purchasing of lamps is simulated. In this model, consumers are modelled based on heterogeneous preferences, and they develop opinions (based on memories and perceptions) about lamps and share these in a social network structure. Lighting technology is modelled using lamps of many different technology types. The results of the simulations indicate that the ban on bulbs will be effective in realising an energy-efficient sector, albeit at significant expense to consumers. An alternative policy, introducing a tax on incandescent lamps, is also shown to be effective, given a sufficient level of taxation.

6.1 Introduction

Lighting is essential for modern living: it enables humans to do many things otherwise impossible, for both work and leisure. Whereas humanity has used artificial lighting for millennia, the last two centuries have seen dramatic increases in the use of lighting. From candles of medieval times candles to today's highly efficient gas discharge and solid state lamps, lighting technology has progressed greatly, contributing to a significant decline in the cost of lighting (Fouquet and Pearson 2006).

Electric lighting really took off after 1879, when Thomas Edison demonstrated his durable, well-performing incandescent light bulb, by using it to light his Menlo Park laboratory complex (NPS 2007). During the last decades of the 19th century, electric power stations were erected in major cities around the world, supplying enough current for up to a thousand incandescent glow-lamps per electric station (Forbes 1889), marking the beginning of the electric power infrastructure.

Edison's first carbon filament glow bulb gave 2 lumens of light per watt of electricity and boasted a lifetime of 45 hours (Gendre 2003). Many gradual improvements increased the lifetime and efficiency of the bulbs. By 1912, the glow bulb's

E.J.L. Chappin (✉)
http://chappin.com/emile
e-mail: e.j.l.chappin@tudelft.nl

K.H. van Dam et al. (eds.), *Agent-Based Modelling of Socio-Technical Systems*, Agent-Based Social Systems 9,
DOI 10.1007/978-94-007-4933-7_6, © Springer Science+Business Media Dordrecht 2013

efficiency had improved to reach a light output of 12 lumens per watt of electricity. But since then, technological progress has been more or less stagnant. Presently, almost 100 years later, the incandescent lamp is hardly more efficient: circa 98 % of the electricity used is converted into heat and not light.[1]

More energy-efficient alternatives to the incandescent lamp have been developed, for example the compact fluorescent lamp (CFL) and the light-emitting diode (LED) (Azevedo et al. 2009). The compact fluorescent lamp was first introduced by Philips in 1980. It offered 4 times the energy savings of incandescent lighting and had a much longer lifetime, but there were disadvantages regarding size, startup time and weight. The CFL has since been improved and has become popularly known as the *saving lamp*. The CFL enables a dramatic increase in the energy efficiency of lighting. It also retains an amount of compatibility with existing luminaires.[2] CFLs offer clear benefits for many applications. Many governments have tried to stimulate their use (see e.g. Mills 1993; Martinot and Borg 1998), but these stimulus programs have had only limited success. Today, CFL saving bulbs are present in only 55 % of European households (Bertoldi and Atanasiu 2007).

Although LED technology has existed for almost half a century, it is new in the area of consumer lighting. General Electric introduced the first commercial LEDs in 1962 (Azevedo et al. 2009) and initially they were only available in the colour red. Since then, the developments in LED technology have continued. LED lamps are now a very promising energy-extension alternative. In the laboratory, LED designs achieve unparallelled electric efficiencies compared to other light sources (Dupuis and Krames 2008). Proponents consider the LED to be the ultimate lamp of the future, the main reasons being its suitability to a wide range of applications and its potential to gain in electric efficiency (Azevedo et al. 2009; Curtis 2005; Holonyak 2005; U.S. Department of Energy 2009).

Despite the availability of these new types of lamps, the inefficient incandescent lamp is the type still predominantly used by households. Consequently, much electricity is wasted in the consumer lighting sector. For the Netherlands alone, yearly electricity usage in consumer lighting now equals about 3.8 TW h_e. This is comparable to the output of one large coal power plant (800 MW_e). If 40 % of this could be saved by switching to more efficient lighting (from Mills 2002), this would equate to removing a 320 MW_e plant from the grid (Afman 2010).

There are a number of reasons as to why consumers only reluctantly adopt energy-efficient lighting options (Menanteau and Lefebvre 2000). Both CFLs and LEDs are characterised by high up-front costs. This forms a significant barrier for adoption, because consumers typically factor in high discount rates when purchasing energy-efficient durable goods (Hausman 1979; Kooreman 1996). Yet consumers' knowledge about present designs may be outdated. Perceptions about the performance of CFLs are known to be more negative than actual recent numbers (see e.g. Martinot and Borg 1998; Peifer 2007).

---

[1] based on an efficiency of 12 lm/W and a theoretical maximum of 683 lm/W (Azevedo et al. 2009).

[2] *Luminaire* is the technical term for a lighting fixture. A luminaire is the complete unit, including lamp, and, if applicable, reflector, ballast, socket, wiring, diffuser and housing (Sylvania 2009).

However, in consumer lighting, changes are forthcoming. The European Union's phase-out of incandescent lighting is a clear strategy that will change the sector. It involves regulation designed to remove from stores the cheapest forms of inefficient household lighting CEC (2009). Nevertheless, it is uncertain whether the lighting sector will become efficient overnight. Consumers may switch to forms of inefficient lighting that are exempt from the phase-out. We conjecture that simulation will prove useful in understanding the dynamics induced by the phase-out.

In the rest of this chapter, we will describe the agent-based model in the ten steps that were discussed in Chap. 3.

6.2 Step 1: Problem Formulation and Actor Identification

The main question addressed in this chapter is whether it is likely that the EU ban on bulbs will bring about a transition to low electricity consumption in the Dutch consumer lighting system. We have focused on the Dutch market, although many aspects in consumer lighting are more generic. The focus on the Dutch market allowed for an operational model, with country-specific data. In principle, the same model can be used to simulate many other countries, simply by using different data sets. Stakeholders are consumers, retailers of lamps and lighting stores, and manufacturers of light bulbs and lighting fixture designs. Government actors are also relevant stakeholders: specific to the Dutch situation, we may point to the task force on lighting installed by former minister Cramer of Housing, Spatial Planning and the Environment (Taskforce Verlichting 2008). The Ministry of the Environment (VROM) is the problem owner. A consumer markets specialist was asked to cooperate as domain expert in the development of the model and the evaluation of the results.

6.3 Step 2: System Identification and Decomposition

The task at hand in the system identification and decomposition step is to come to a systems description suitable for the design of the model. In this section we will describe this process.

6.3.1 Inventory

The outline of the consumer lighting sector was established by applying *theory* to, among other things, complex systems and consumer markets. In addition, literature on consumer lighting and related consumer markets was analysed. It followed from the analysis that the consumer lighting sector can be considered a *complex socio-technical system*, comprising:

- A social subsystem that consists of social actors, with their behaviours, interactions and developments;
- A technological subsystem that consists of the different lighting 'hardware' objects, with their physical properties, technological intricacies, compatibilities and developments; and
- Various interactions within and between these subsystems.

These will be described below.

6.3.1.1 Social Subsystem

The actors in the social subsystem are represented as *agents* in the model. The main actor in the consumer lighting system is the *consumer* itself. Consumers take the actual decisions about which light bulb or luminaire to purchase and when to switch on or off each of their lights. Therefore, the consumer is at the core of the model. A consumer, in this case, is the representation of a household.

Consumer choice is narrowed by a variety of factors. *Manufacturers & developers* are key players in innovation and in the selection of the products for the market. *Lighting retailers* then sell only a selection of the total available products to their customers. Finally, the government sets the rules of the game: it implements policies and acts on them, for instance by collecting taxes.

6.3.1.2 Technological Subsystem

In addition to the agents, which represent actors in the consumer lighting system, there are technological components, representing physical objects. The following list comprises the main components in the physical subsystem:

- The range of all the different kinds of *lamps*, or light bulbs,[3] used by households.
- The different *luminaires*, or light fixtures, used in homes. Luminaires can be free-standing or fixed to light points in the house.
- The light *points* in the house are the fixed outlets in ceilings or walls. Light points also include other locations where people plug in lamps, such as a desk or locations for standing lamps.
- Light *switches* including dimmers allow consumers to control lamps.
- Other apparatus are used to couple lamps to the mains. Examples are electromagnetic and electronic transformers, CFL ballasts, current control modules, electric wiring, power strips, extension cords and fuses.

This list implies that the physical system is demarcated to include the physical components from the power outlet on to where the light is emitted. This demarcation is shown in Fig. 6.1.

---

[3] A *lamp* is a broad term reflecting a source of man-made light. A lamp can refer to a light bulb or an the entire device, such as oil lamp, gas lamp (Dictionary.com 2010). In this chapter, we consider lamp and light bulb as synonyms.

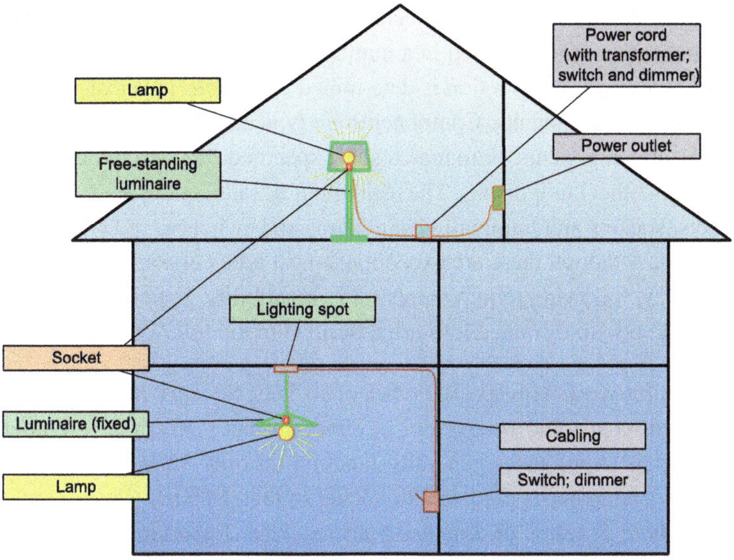

Fig. 6.1 Overview of the main technological components in consumer lighting

6.3.1.3 Interactions

Interactions take place between the components. We define social interactions to be between social components, technical interactions to be between technical components, and socio-technical interactions to be between social and technical components.

Social Interactions Social interactions are communications and negotiations between social elements or agents. A main type of social interaction is contact between consumers through their social network. Consumers may discuss lighting decisions and experiences with others or observe others. Social interactions of this type are called *word-of-mouth*. These interactions generally affect perceptions, which are input into decisions regarding the purchase of lamps. Eventually, word-of-mouth may impact the electricity consumption for lighting.

Government interacts with consumers by providing information through information and awareness campaigns. Furthermore, government can implement policies, such as bans on products, taxation schemes and subsidies. Consumers interact with retailers by visiting their stores, collecting information and purchasing light bulbs and luminaires. Government may aim to influence manufacturers when their targets require more energy-efficient lamps. Businesses—manufacturers and retail stores—also interact in order to entice the consumer into purchasing products by using marketing strategies such as promotions.

Technological Interactions The technical subsystem consists of an assemblage of components that are connected in a number of ways and hence interact. The nature of a technological interaction is determined by various physical and electronic properties of the components. Components are typically designed to be compatible with a class of components, or to match some specified standard in order to interact optimally with other components. The technological interactions are between: lamps and sockets, sockets and luminaries, luminaires and switches, and switches and the power source. Although there are exceptions, most types of components connect in a standard way, following requirements for compatibility. Important characteristics that determine possible connections are described in the following list:

- All luminaires work with specific types of sockets. Sockets, in turn, are designed to work with a specific voltage: 12 V direct current (DC) or 230 V alternating current (AC).[4] Appliances generally function on only 12 V DC or 230 V AC. Therefore, a lamp needs to match the socket so that the lamp-fixture combination functions both in terms of dimension and voltage. Important are E27 and E14, both 230 V screw sockets, which have been traditionally used for incandescent bulbs, E27 being larger then E14. CFLs and some LEDs also fit in these sockets. More recently, sockets for halogen lamps have been introduced. GU10 is a 230 V click-and-turn socket used for spots, G53 a socket for 12 V spots, and R7S a 230 V socket with a tube shape.
- A luminaire can be free-standing and therefore autonomous in the sense that the only coupling is through a 230 V power outlet. Furthermore, a luminaire in a fixed location in the house often is connected to one or more switches that allows interruption of the connection.
- Switches traditionally operate by connecting or interrupting the electricity connection to a luminaire. More recently, switches have been developed that contain electronics compatible with only some specific lamps and luminaires. An example of such switches is a dimmer. There are different types of electro-magnetic or electronic dimmers that use touch, some memorise their last state. As some lamps cannot be dimmed, such switches cannot always be used. Furthermore, lamps themselves react in different ways to dimmers, causing, for example, a need for reactive power.
- A luminaire can contain electronics that are specific to certain types of lamps. Examples are the electromagnetic or electronic ballast and starter modules for a fluorescent light tube, an external ballast for some CFLs, a transformer for a 12 V halogen, and a driver for LED arrays.

[4] A variety of standards exist for the voltage of consumer appliances. We focus in this chapter on 230 V AC, which is common in Europe.

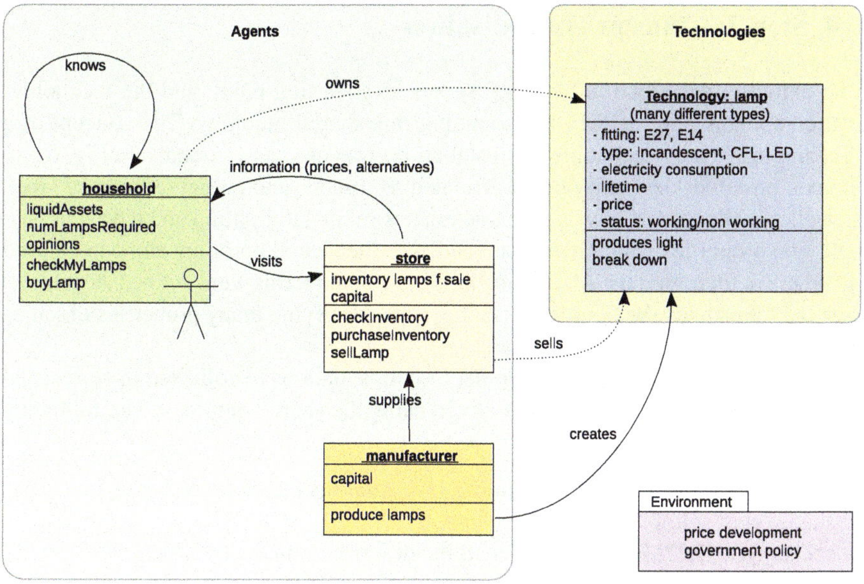

Fig. 6.2 Structured overview of components and interactions in the consumer lighting system

Socio-technical Interactions There are also interactions between social system components and technology. The first type of socio-technical interactions is when social actors need to make purchase decisions regarding lamp products. Actors may not be able to understand the intricacies of the technology involved. The technological characteristics and requirements may be difficult to understand.

The operation of lamps is another type of socio-technical interaction. Other socio-technological interactions arise due to improvements in energy-efficient (CFL/LED) technology. These technological developments result in political processes. For instance, the EU ban on bulbs is only possible because a range of alternatives are available that may replace incandescent bulbs.

6.3.2 *Structuring*

A structured set of components and interactions was described above, which was the result of brainstorms. One way to describe the *structure* in these components and interactions is to visualise their connections. Of the technical system, that was done in Fig. 6.1. The structure of the consumer lighting system as a whole is depicted in Fig. 6.2.

6.4 Step 3: Concept Formalisation

The ontology described in Sect. 4.2 serves as a starting point, and the technological components are defined by extending this shared ontology. The concept of a 'Technology' (i.e. a technological installation) was used and extended in Fig. 6.3 to include possible 'design properties' related to 'brand' and energy-efficiency label, as well as a number of physical characteristics related to lighting and electricity usage. The properties which have been added to the shared ontology and can be used by others in their models. Also, some lighting-related terms were defined as a 'label' (see the far right of the figure). This allows for specifying many properties of lamps and luminaires within the ontology.

In order to compile an extensive list of lamps, data were collected in stores, and the instances of these lamps were defined using the shared ontology. The following properties for lamps were set:

- technology type: CFL, LED, halogen or incandescent;
- shape type: spiral, tubular, sphere or spot;
- socket type: E27, E14, GU10 (recall the description in Sect. 6.3.1.3);
- brand name: Ikea, Philips, etc.;
- design lifetime: lifetime mentioned on the package, in hours;
- electricity consumption: power in wattage;
- quantity of light output: number of lumens;
- colour of light: measured in colour temperature (K), and colour rendering index; and
- status: operational or failed.

The luminaire contains the following properties:

- list of fitted lamps;
- location in the house;
- usage (hours per week);
- applicable socket, number of sockets; and
- maximum power (W).

The social part of the system consists of all agents that represent actors. They are displayed in the far left of Fig. 6.3. The different types of agents (i.e. household, retailers and government) are all extensions of a generic Agent defined in the ontology. The retail store is formalised to be integrated with the manufacturer. It has an inventory of lamps and luminaires that are for sale and can produce as many lamps as are being sold. The government has a set of policies which can be activated, based on the policy scenario that is run.

Some parts of the concept formalisation were not entered in the shared ontology for socio-technical systems (see the boundary line in Fig. 6.3) because they are too specific for this single model to be widely re-used or because they are difficult to fit in the structure of the existing ontology. The agents, despite their general names, have specific behaviour and properties which would make them incompatible with other agents of the same name in other models. Furthermore, even though a

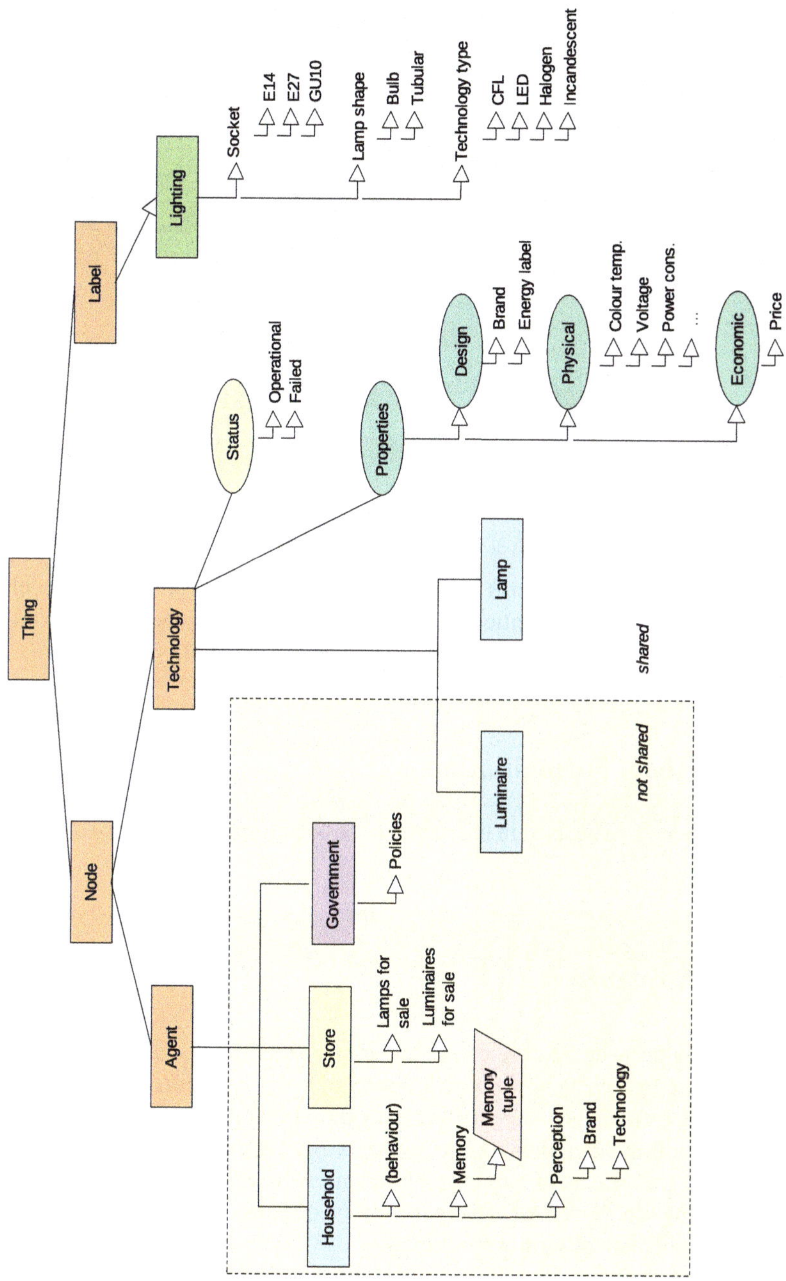

Fig. 6.3 Extension of the shared ontology for socio-technical systems with case-specific concepts

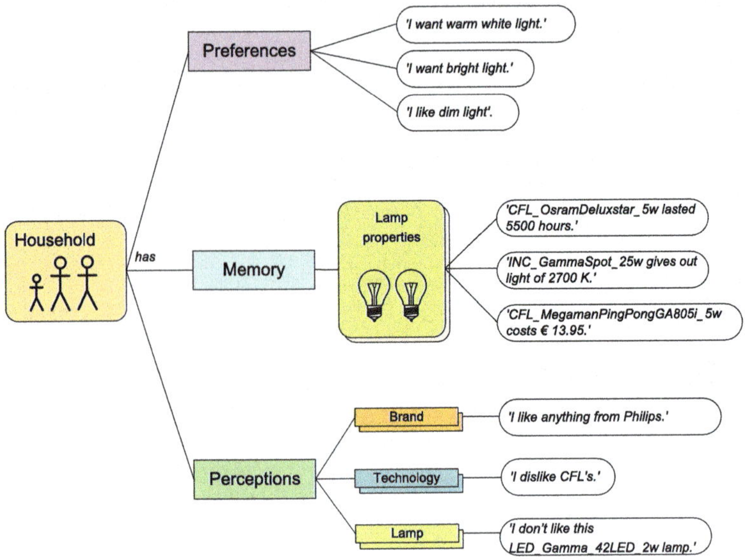

Fig. 6.4 Conceptual view of a household's data structure concerning lamps

lumanaire fitting a number of lamps is not incompatible with the ontology, it would have required a very model-specific extension and hence it was deemed not worth including it in the more generic shared ontology.

6.5 Step 4: Model Formalisation

Now that the concepts to be used in the model have been formalised, the model itself can be set up.

6.5.1 Social Structure

The household agent is at the core of the model: it makes the decision to purchase lamps. In deciding which lamp to buy, a household will need information on the different existing alternatives. Agents use their individual preferences to compare the alternatives and make a choice. The household is equipped with a set of preferences for different aspects of lamps and luminaires. In addition, the agent retains a memory for lamp characteristics, for perceptions of lamps, technology types and brands. This is schematically shown in Fig. 6.4.

Household agents are incorporated in a social network structure. Links in the social network reflect that agents know each other. The households have perceptions that are shared over this network structure. Two network generation algorithms were

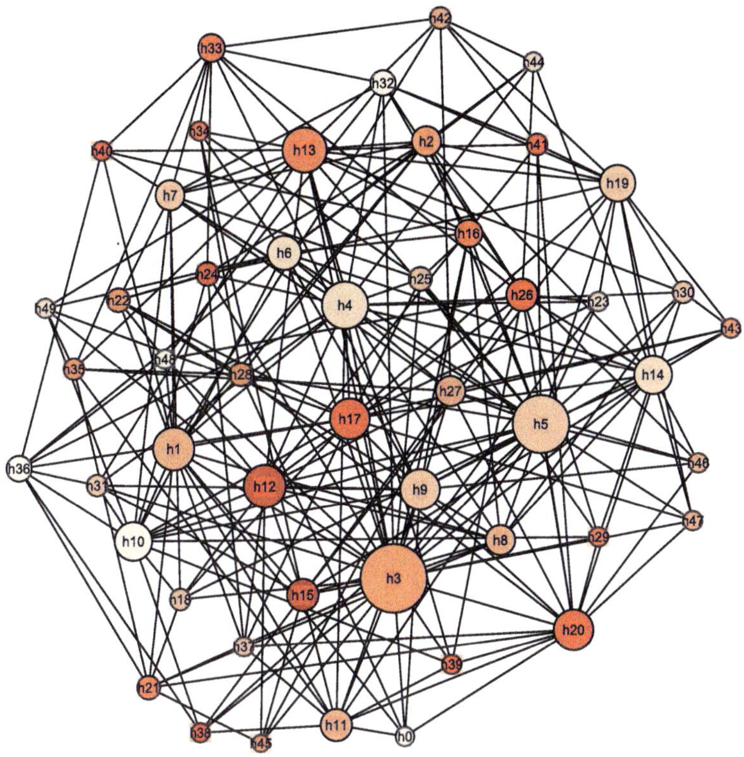

Fig. 6.5 Scale free network of households, for 50 households, with degree parameter $k = 6$. The size of the node corresponds to the node degree; the *colours* differ according to the household lamps' energy efficiency

implemented: the small-world network according to the Watts and Strogatz algorithm (Wang and Chen 2003) and the scale-free network according to the growth and preferential attachment mechanisms (Barabási and Albert 1999). The scale-free network structure was used because it was deemed most proper for rumour spreading in social networks (Nekovee et al. 2007). The implementation of these social networks was done in a generic way, so they can be used in other models with consumers as well. An example of a scale-free network is displayed in Fig. 6.5.

At the initialisation of a simulation, 250 consumers are generated, each with its own preferences. They are placed in a scale-free social network in which they know 16 other agents on average. Each household starts with a set of lamps. As a whole, the agents represent the Dutch consumers in terms of the distribution of lamp technologies and numbers of lamps per household.

6.5.2 Model Narrative and Pseudo Code

During the simulation, consumers fulfil their needs for lighting. As time passes, lamps are used and break when their technical lifetime has expired. Consumers

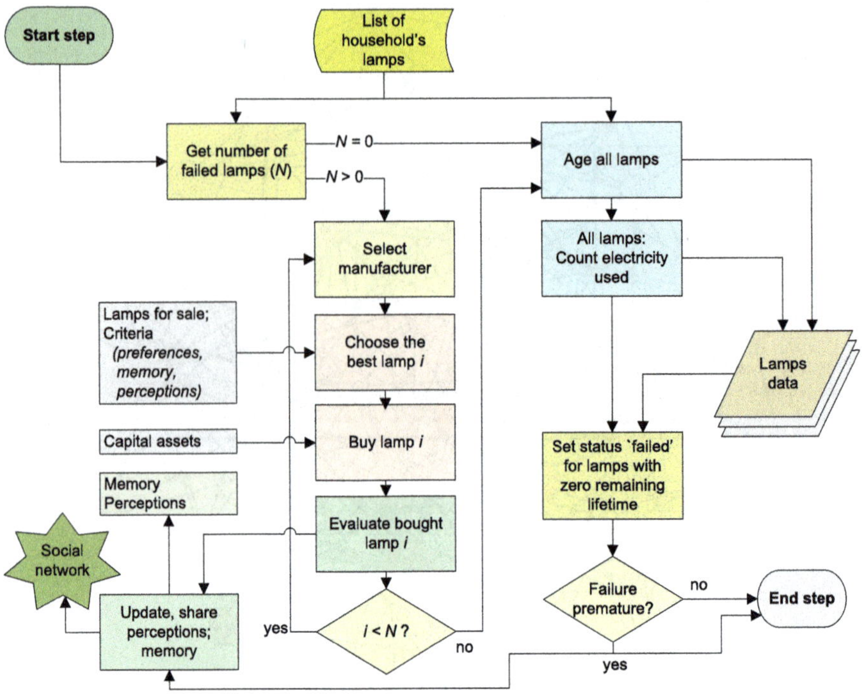

Fig. 6.6 Flow chart representing the sequence of actions during the household's step

discard and replace broken lamps by visiting the store and buy the lamp of their preference. Over time, these purchase decisions are affected by price changes, government policies, experiences with lamps and shared perceptions. The activities of households are structured in a sequence of actions, which is described below.

A simulation tick represents one week and consists of all households performing a sequence of actions, one after another. A household performs the following actions (see Fig. 6.6):

- The household checks if any of its lamps have failed by looking at their status.
- The household replaces broken lamps. For this the household visits a retail store. Deciding which lamp to choose is done by means of a multi-criteria analysis (which was developed in a generic manner for the emissions policy model in Chap. 7). In the multi-criteria analysis, the purchase decision consists of comparing the alternatives on a number of criteria, which have agent-specific weights attached to them. The criteria are household's preferences, perceptions and knowledge on lamp aspects.
- When the household has bought new lamps, they are evaluated with respect to characteristics that are not on the package but are relevant. This updates the household's memory and perceptions. The perceptions are shared with a random other agent from the household's social network.
- The household uses its lamps for a duration specific to the location of the lamp. Use causes lamps to age. The consumed electricity is recorded.

- When the remaining lifetime of a lamp of a household is equal to or lower than zero, the lamp's status is changed to 'failed'. In the next time step it will be replaced. If the household notes that a lamps has failed prematurely—when the actual lifetime was much shorter then the lifetime promised on the package—the households' perception of this lamp type decreases. Again, these perceptions are shared in the network.

As an example, the second step (replacement of broken lamps) is described in pseudo-code in Algorithm 6.1. Other tasks can be translated from the model narrative into pseudo-code in a similar fashion, before implementing the model in Step 5.

Algorithm 6.1 Replacement of broken lamps

```
for each consumer
  for each lamp that is marked as 'failed'
    select manufacturer
    select lamps for sale
    choose the best lamp according to criteria regarding preferences, memory and perceptions
    buy new lamp
    evaluate the bought lamp
    update perceptions of lamp technology, lamp model, and lamp brand based on the new purchase
  end for

  for each lamp
    use and age all lamps
    if lamp lifetime passed
      mark lamp as 'failed'
      if failure is premature
        update perceptions of lamp technology, lamp model, and lamp brand based on this failure
      end if
    end if
  end for
end for
```

6.6 Step 5: Software Implementation

The model was implemented using standard practices as outlined in this volume. It was programmed in Java, containing some 4500 source lines of code. The code contains a class for the model itself and a class for each of the types of agents. Also classes for 'Lamp' and 'Luminaire' were implemented, which have no behaviour but mainly store some fields and information.

All the lamps used in the model were stored in the shared ontology. Many shared Java classes were used, such as those for agent, technology, multi-criteria analysis and contracts. Data from simulations were stored in a database and analysed with Matlab scripts. Graphs were created showing the model output for a number of indicators. Work was performed to streamline the process of running models in the distributed computation system.

6.7 Step 6: Model Verification

In order to verify the model several tests were used. These included a range of structure-behaviour tests that focused on the outcomes of purchase decisions by consumer agents. The range tested comprised a set of model runs that completed a parameter sweep of 25 model parameters. Over 10 million purchase decisions were analysed. It was verified whether these were as expected, based on the logic behind the model. The data was analysed by making pie charts of slices of the data. For instance, from the result of purchasing a lamp using only one criterion it is easy to deduce whether that criterion works as expected and as it should be. By systematically analysing individual criteria and combinations, the purchase decision in the model was verified.

6.8 Step 7: Experimentation

The model was run in an experimental setup of 1600 runs comprising of four scenarios. We modelled a base case with no governmental policy and three cases with different possible policy interventions. The first policy was a ban on incandescent bulbs. The second policy was a taxation scheme on standard light bulbs. Finally, a subsidy scheme on LED lamps was modelled. These policy variables are formalised as follows:

Ban on bulbs — This policy experiment entails a complete ban on the standard incandescent light bulb, phased in between year 2 and 5 from the start of the simulation. This policy is comparable to the EU ban on household light bulbs. First the incandescent bulbs with the highest wattages are removed from the stores, after which the lower wattages are progressively removed.

Incandescent taxation — This policy scenario introduces a taxation on the sale of incandescent light bulbs. This tax increases progressively during the first five years of the simulation to €2.00 per lamp. This tax is relatively large, as an incandescent lamp costs between €0.35 and €1.50.

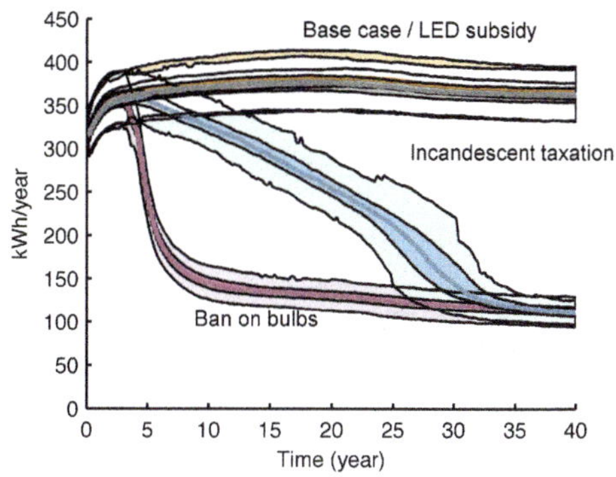

Fig. 6.7 The average electricity intensity (electricity used for lighting) of the modelled households (kWh/year)

LED subsidy — The third policy is a subsidy on the purchase price of LED lamps. The subsidy is a discount of 33 % of the purchase price. The LED subsidy starts after year six and lasts for five years.

For each policy case, the simulation is executed 100 times in order to be able to explore the spread between runs. Furthermore, two key model parameters regarding criteria weight factors are varied. The weight factor for price is either 4 or 6 on average. The weight factor for normative influence is either 2 or 4 on average.

6.9 Step 8: Data Analysis

In order to analyse results from the simulations, graphs were made of the main indicators over time. The main indicator for the success of a government policy was the average electricity consumed by households for lighting (see Fig. 6.7). Underlying this electricity use were the levels of adoption of the different lamp technology types (see Fig. 6.8). The money spent on lamps purchases was also analysed. Other parameters were used to increase understanding of how the figures for these indicators came about. These included perceptions of the different lamp models, technology types and brands.

From a single run of a simulation experiment it is not possible to draw substantial conclusions. Values in parameters will vary between runs because of the use of stochastics during agent initialisation and model execution. Stochastics were therefore used to make agents heterogeneous and to randomise lamp failure. To be able to arrive at a realistic assessment of patterns observed in the simulated system's evolution, one needs to look at the *average* and *spread* of the results of many runs. We are interested in the spread of the outcomes around the average and their skewness, so we used the concept of a so-called *box plot*. In a box plot the following statistics populate the box and "whiskers":

- Mean;
- Inner quartile range (50 % of the data);

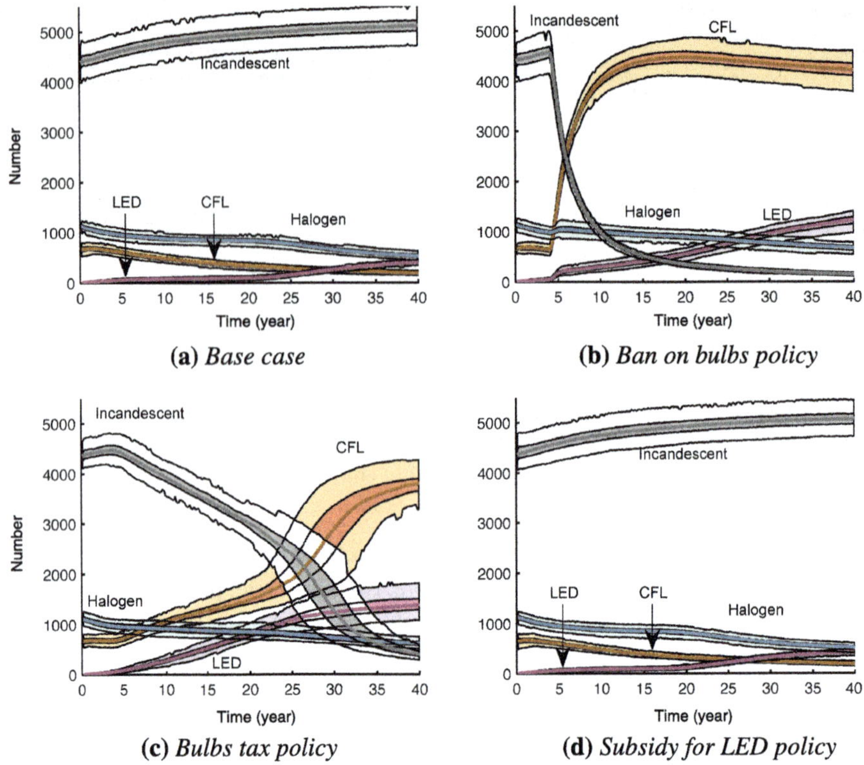

Fig. 6.8 The total number of lamps in the simulation, for each of the lamp types

- Lower and upper "whisker lines" (for data following a normal distribution, 98.6 % of the data); and
- Outliers (if any / if significant).

Plotting these statistics for all data points on the horizontal axis gives us a graph that is useful in showing both the mean and the spread of outcomes of an experiment. The main indicator for success is the average electricity intensity of households. If our aim is to achieve maximal energy savings in lighting, this is the most important indicator.

Results for the electricity intensity are plotted in Fig. 6.7. Both the ban on bulbs and the tax case are likely to be effective at reducing the electricity intensity. As said, underlying this electricity use is the composition of the technology portfolio of lamps in use, which is displayed in Fig. 6.8.

An example of when it is important to include outliers in the analysis is shown in Fig. 6.9. For these runs the settings have been tweaked specifically to demonstrate the effect of such outliers. What can be seen is that in some runs halogen technology becomes dominant, and in others, CFL technology. These are very different states but are caused by different random numbers. The eventual dominant technology in this case depends the preferences of the first few. Please note that the results are

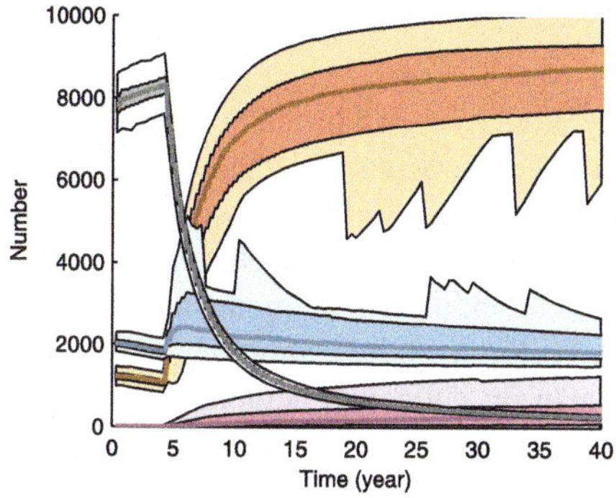

(a) *Ban on bulbs policy, no outliers*

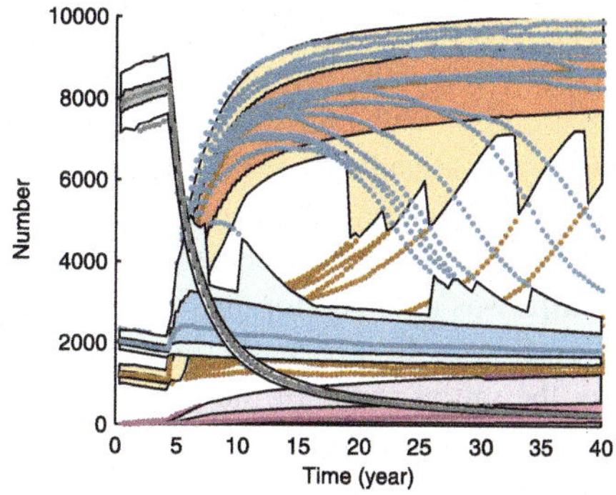

(b) *Ban on bulbs policy, outliers shown*

Fig. 6.9 The total number of lamps in the simulation, with different simulation settings, for about 40 runs per experiment. Outliers are plotted in the graph (**b**) at *right*. This reveals quite a different underlying pattern for a number of runs from what is visible in graph (**a**), at *left*. Halogen (shown in *blue*) is decidedly more popular than CFLs in a about 7 runs, whereas it was less popular in the other runs

plotted for the adoption of lamp types, with and without outliers, for the policy scenario of the ban on bulbs.

6.10 Step 9: Model Validation

The main task in the validation process is to show that the model is *fit for its purpose*. As we are exploring transitions in the energy domain, specifically transitions in consumer lighting, the question is whether we have created a model that helps us to adequately address this question. We do not claim that the model allows us to accurately predict such a transition or the actions taken by companies, governments, shops, manufacturers and consumers. Rather, we intend to compare policies regarding consumer lighting to allow for a better informed choice between these. In this light, a number of validation techniques were used.

A main source for validation is the use of existing theory for developing the agents. The consumer agent was based on theory of consumer behaviour. Input from experts was used. In part, the survey data could be used to feed the consumer purchase decisions. Additional specific experimental data was gathered at the time of writing to further underpin the consumer decision algorithm.

Finally, an additional experiment was performed for the purpose of validation. In that experiment, we aimed to replay patterns found in the history of the consumer lighting sector from 1980 onwards. That required a number of different settings—the starting portfolio, change in luminaire, and older lamps—but was rather successful (Chappin 2011).

6.11 Step 10: Model Use

The model results have been used in policy advice regarding the consumer lighting sector. Results have been conveyed to the Dutch Ministry of the Environment and several interested institutes (e.g. Agentschap NL), as well as in conferences and workshops with modellers and people from the lighting industry. During the study, Philips was involved for the purpose of providing input into the model as well as to learn from it from a marketing perspective. Philips withdrew, however, when during the initial phase of the project the EU ban on bulbs was actually announced. Philips argued at the time that the model was unnecessary because the policy decision had been made. In addition to the comparison with other options, the results show possible drawbacks and inefficiencies with the current policy. The most important example is the role of halogen lamps: although they are not more efficient than incandescents they are not phased out by the ban. Given the popularity of halogens over the past decade, the sector may shift to a suboptimal state in terms of energy efficiency. Our simulations show that this may indeed happen.

Furthermore, additional experiments could be specifically designed from a marketing perspective. Those experiments could by very insightful for companies such as Philips: which marketing strategies are likely to be successful? One could consider, for instance, whether a company should wait with the introduction of LEDs until they perform well or start the introduction as soon as possible. Such experiments could be performed with the model with minor adaptions in the way new lamps are introduced. Similarly, many other interesting experiments could be formulated.

6.12 Conclusions

From the simulations, we conclude that it is likely that the EU's ban on bulbs policy is an effective way to curb the use of incandescent lamps for consumer lighting. The adoption of incandescents will likely decline as purchases of more energy-efficient lamps increase. The ban on incandescent bulbs will therefore rapidly reduce electricity consumption in the consumer lighting sector. The bulbs-tax policy would also be effective in reducing the use of the incandescent lamp and decreasing household electricity consumption. However, it might well take longer to reach similar consumption levels. The prevalence of halogen bulbs in some runs is a cause for concern: some halogens are not more efficient than the incandescent lamps they replace. In contrast to the other policies, the policy to subsidise LEDs is unlikely to affect consumer decisions.

In future research, a reflection on the robustness of the results can be done by assessing different levels of heterogeneity of households. Furthermore, classes of consumers well known in marketing literature may be identified explicitly (e.g. early adopters, late majority and so on). These classes could be made explicit, incorporating a discount rate people use in assessing financial benefits of investing in saving lamps. A number of rebound effects of government policies can be added to the simulation model. One rebound effect would be that people are more inclined to keep their energy-efficient lamps burning for a longer duration.

In this chapter it was demonstrated how a model of the lighting sector, including consumer behaviour, can support decision-making on policies to influence the electricity consumption of the residential sector. This model is based on a survey of lighting consumers, data from lamps in stores and literature on consumer decisions. Based on these things, we chose to model consumer behaviour by using preferences, perceptions, memory, and multi-criteria decision-making. Consumers in the model engage in information sharing and normative adaptation because they fit in a social network that is generated at the start of the simulations.

References

Afman, M. (2010). *Modelling transitions in consumer lighting—consequences of the E.U. ban on light bulbs*. Master's thesis, Delft University of Technology, Faculty of Technology, Policy and Management.

Azevedo, I. L., Morgan, G., & Morgan, F. (2009). The transition to solid-state lighting. *Proceedings of the IEEE*, *97*(3), 481–510.

Barabási, A.-L., & Albert, R. (1999). Emergence of scaling in random networks. *Science*, *286*(5439), 509–512.

Bertoldi, P., & Atanasiu, B. (2007). *Electricity consumption and efficiency trends in the enlarged European Union—status report 2006*.

CEC (2009). Directive 2008/101/ec of the European Parliament and of the Council of 19 November 2008 amending directive 2003/87/ec so as to include aviation activities in the scheme for greenhouse gas emission allowance trading within the community. *Official Journal of the European Union*, *8*(L076), 3–21.

Chappin, E. J. L. (2011). *Simulating energy transitions*. PhD thesis, Delft University of Technology, Delft, the Netherlands. ISBN: 978-90-79787-30-2.

Curtis, S. (2005). Efficiency gains boost high-power LED performance. *Compound Semiconductor*, *11*(11), 27–30.

Dictionary.com (2010). Lamp, dictionary.com unabridged. http://dictionary.reference.com/browse/lamp.

Dupuis, R., & Krames, M. (2008). History, development, and applications of high-brightness visible light-emitting diodes. *Journal of Lightwave Technology*, *26*(9), 1154–1171.

Forbes, G. (1889). Electric-lighting stations in Europe, and their lessons. *Science*, *13*(326), 337.

Fouquet, R., & Pearson, P. J. G. (2006). Seven centuries of energy services: the price and use of light in the United Kingdom (1300–2000). *Energy Journal*, *27*(1), 139 177.

Gendre, M. F. (2003). Two centuries of electric light source innovations. URL: http://www.einlightred.tue.nl/lightsources/history/light_history.pdf.

Hausman, J. A. (1979). Individual discount rates and the purchase and utilization of energy-using durables. *The Bell Journal of Economics*, *10*(1), 33–54.

Holonyak, J. N. (2005). From transistors to lasers and light-emitting diodes. *Materials Research Society Bulletin*, *30*, 509–515.

Kooreman, P. (1996). Individual discounting, energy conservation, and household demand for lighting. *Resource and Energy Economics*, *18*(1), 103–114.

Martinot, E., & Borg, N. (1998). Energy-efficient lighting programs. Experience and lessons from eight countries. *Energy Policy*, *26*(14), 1071–1081.

Menanteau, P., & Lefebvre, H. (2000). Competing technologies and the diffusion of innovations: the emergence of energy-efficient lamps in the residential sector. *Research Policy*, *29*(3), 375–389.

Mills, E. (1993). Efficient lighting programs in Europe: cost effectiveness, consumer response, and market dynamics. *Energy*, *18*(2), 131–144. IN1.

Mills, E. (2002). The $230-billion global lighting energy bill. In *Proceedings of the fifth European conference on energy-efficient lighting* (pp. 368–385).

Nekovee, M., Moreno, Y., Bianconi, G., & Marsili, M. (2007). Theory of rumour spreading in complex social networks. *Physica A: Statistical Mechanics and Its Applications*, *374*(1), 457–470.

NPS (2007). Edison biography. Website. Accessed: May 19th, 2009. URL: http://www.nps.gov/edis/historyculture/edison-biography.htm.

Peifer, D. (2007). The CFL myth. *Lighting Design and Application LD and A*, *37*(10), 87–90.

Sylvania (2009). Business products: Glossary, Website. Accessed: November 2009. http://www.sylvania.com/BusinessProducts/Glossary/.

Taskforce Verlichting (2008). *Groen licht voor energiebesparing—eindrapport van de taskforce verlichting*. Report, SenterNovem.

U.S. Department of Energy (2009). *Multi-year program plan fy'09-fy'15—solid-state lighting research and development*. Technical report, prepared for: lighting research and development, building technologies program, office of energy efficiency and renewable energy. U.S. Department of Energy, Prepared by: Navigant Consulting, Inc., Radcliffe Advisors, Inc., and SSLS, Inc.

Wang, X. F., & Chen, G. (2003). Complex networks: small-world, scale-free and beyond. *IEEE Circuits and Systems Magazine*, *3*(1), 6–20.

Chapter 7
An Agent-Based Model of CO_2 Policies and Electricity Generation

E.J.L. Chappin and G.P.J. Dijkema

Abstract The electricity infrastructure is essential to the functioning of society. In this chapter we describe the way in which we have modelled this socio-technical system in order to analyse CO_2 policy instruments, comparing the results of "no intervention" with those of implementing taxation and emissions trading to reduce CO_2 emissions in the production of electric power. Because power plants currently in operation have only a small emissions reduction potential, a reduction of 20 % by 2020 or 50 % by 2050 would require all new units to be designed and constructed to have near-zero CO_2 emissions. In the model, power production companies own, operate and invest in power plants. They sell electricity and buy fuels and emission rights in simulated markets. When demand grows or assets age beyond their technical lifespan, they must decide on investments and select their technology of preference.

7.1 Introduction

In the European Union, the United States and countries around the globe, electric power production has been transformed from a state-controlled sector or monopoly to a true, contestable market in which private companies compete. Many actors own, develop and operate parts of this socio-technical system, and its characteristics and evolution emerge as a result of the interaction within and between the social and the technical (Dijkema and Basson 2009).

Electric power production accounts for 36 % of the CO_2 emissions of the EU.[1] There is a growing consensus that CO_2 emissions need to be stabilised and then reduced significantly in the course of this century to avoid global warming of more than 2 °C (IPCC 2007). Using an economic term, CO_2 is a *negative external effect* of power production. It is disposed of in the atmosphere, which is a 'common-pool resource' (Hardin 1968). To internalise this external effect, command-and-

[1] 2007 data based on http://ec.europa.eu/energy/publications/statistics/statistics_en.htm.

E.J.L. Chappin (✉)
http:chappin.com/emile
e-mail: e.j.l.chappin@tudelft.nl

K.H. van Dam et al. (eds.), *Agent-Based Modelling of Socio-Technical Systems*, Agent-Based Social Systems 9,
DOI 10.1007/978-94-007-4933-7_7, © Springer Science+Business Media Dordrecht 2013

control regulation has traditionally been the type of regulation employed: environmental laws have been developed, implemented and enforced (cf. McCormick 2001). Inspired by the success of SO_2 and NO_x trading schemes developed in the USA, much attention has been given to incentive-based instruments in the quest for cost-effective emissions reduction. Whereas command-and-control instruments depend on regulation and legislation, incentive-based policy instruments intend to create market signals that will influence decision-making and behaviour (Egenhofer 2003). The most well-known incentive-based policy instruments that can be used to reduce CO_2 emissions are emissions trading schemes (ETS) and carbon taxation (CT).

By ratifying the Kyoto protocol the EU has agreed to strong CO_2 targets. To achieve these targets the EU has developed its emissions trading scheme, the EU ETS. In retrospect, the scheme was only partly effective in the first phase (2005–2007). Despite adaptations in the current Phase 2 (2008–2012), the price of CO_2 has remained low and production companies continue to plan and build CO_2-intensive coal-fired facilities. It is thus apparent that policy makers face a recurring design challenge for CO_2 policy, namely how to make a policy that works. To help policy makers address this challenge, we have developed the model described in this chapter. To allow us to explore the various policy options, analyse their effectiveness and elucidate their strengths and weaknesses, we formulated the following central modelling question:

What is the preferred type of CO_2 emissions policy?

The focus of this chapter is to describe how we modelled the electric power production system as a socio-technical system to analyse CO_2 policy instruments and developed an agent-based model to allow the comparison of "no intervention" with taxation and with emissions trading. Extensive information on the model and an interpretation of its results can be found in Chappin et al. (2009a, 2009b). This model is one of a series of models on energy transitions (Chappin 2006; Chappin and Dijkema 2009; Escalante 2010) which has led to a framework for modelling *transitions* of energy systems (Chappin 2011; Chappin and Dijkema 2010). In this chapter we will describe the model development process in ten steps, as defined in Chap. 3.

7.2 Step 1: Problem Formulation and Actor Identification

Under the Kyoto Protocol, governments accepted CO_2 reduction targets in order to counter climate change (UNFCCC 1998). In Europe the EU emissions trading scheme (EU ETS) was implemented as of January 2005 (Directive 2003/87/EG CEC 2003). In the EU ETS, companies active in specific sectors must be in the possession of CO_2 emission rights that equal the amount of CO_2 emitted (EnergieNed 2006). Any surplus can be sold; any deficit must be compensated by acquiring rights. Thus a price is put on CO_2 emissions, and this negative external effects is internalised, at least in part.

The EU ETS is a cap-and-trade system, in which the volume of rights issued—the cap—determines their scarcity. It is the responsibility of the EU and its Member States to set the cap and arrange for allocation between Member States and sectors. The result of this process is a cap, preferably set at a level in accordance with the EU 'precautionary principle' (EU Treaty, 1992) and CO_2 reduction targets, and geared to allow the market to generate a price of emission rights that stipulates trade amongst the parties and affects emissions reduction.

Like any emissions trading scheme, the EU ETS is based on the assumption that the *invisible hand of the market* (Smith 1776) would lead to emissions reduction by those who can achieve reduction at the lowest cost (Ehrhart et al. 2003; Svendsen 1999; Svendsen and Vesterdal 2003). However, 'abatement investments remain dependent on an elusive carbon price-signal which has failed to emerge' (Escalante 2010). In contrast to expectations, the first trading period of the EU ETS did not result in 'radical change in the development and use of generation technologies' (Hoffmann 2007). Why did this happen, and what can be expected from the redesigned EU ETS Phase III (CEC 2009)? Because of the experience with the EU ETS, some economists have argued that taxation as a policy instrument is preferable to trading (cf. Grubb and Newberry 2007).

Whatever the outcome of a political debate, once the political process has solidified into clear objectives—such as 20 % emission reduction in 2020, 50 % in 2050—for policy makers the objective of the two pricing mechanisms for CO_2 are the same: to achieve such emissions reduction. To arrive at the core problem formulation and actors involved, it is crucial to realise that for any CO_2 emission reduction to materialise, the electricity infrastructure must change. To initiate and allow such a transition to a low-carbon electricity infrastructure is therefore the responsibility of the policy maker; national and international governments must weave a web of regulations that create a playing field where such a transition can occur. Some "aggregate" policy maker is therefore our problem owner: the European Union, its Member States and of course EU and government officials.

The electricity infrastructure is not owned or controlled by the EU nor by the Member States, but rather is composed of electricity producers, transmission system operators, distribution companies, wholesale and consumer retailers, market platforms, brokers and others. Since we focus on CO_2 and have elucidated that it is through investment in and realisation of an alternative generation portfolio that CO_2 emissions will be significantly reduced (Chappin et al. 2009b), the electricity companies are the most important actors because they design, build, operate and maintain the power generation and CO_2-emitting installations. Together with the grid operated by the TSO, it is their systems that constitute the electricity infrastructure. The other actors in the electricity sectors play an important role with respect to the linkage between power producers and consumers, selection between fuel/asset combinations, and facilitation of the trade in CO_2 rights. However, apart from consumers (who drive demand), their impact on actual CO_2 emission is limited. Here the authors perform the roles of facilitator, modeller and analyst.

7.3 Step 2: System Identification and Decomposition

The system boundary of the system studied is the Dutch power generation system. Adopting a socio-technical systems perspective, this system contains power production companies which own physical assets, i.e. their power plants. These companies are active on the electricity market and are subject to Dutch and European law. These are enforced by the Dutch government. The demand for electricity originates from households, horticulture, large-scale process industry, etcetera. To run their plants, companies must buy fuel and emission rights. Thus, the electricity market is interconnected with the fuel markets and the CO_2 market.

Since it is the objective to explore the effects of "no intervention", of taxation and of trading, the focus of the modelling was to adequately represent this socio-technical system to enable the analysis of the interaction of companies through the fuel, power and carbon markets, the interactions of the various markets and the physical interaction of power generators through the grid. The distinction of the social and the technical is considered important, both for analysis and for modelling.

Figure 7.1 presents an overview of the main system components and their relations. Electric power producers own and operate their generation facilities. They sell power on power markets and to wholesalers. In the case of emissions trading, they are also active on CO_2 markets. Industrial consumers and retail companies buy their power from the market, and wholesalers and households acquire their electricity from retail companies. System operators govern the grids that transport power (a transport network for high voltage and a distribution for medium- and low-voltage transport). The technical infrastructure is a technical system in which fuel is converted to electricity in power plants. The electricity is fed into the grid, which connects producers and consumers. In reality, large industrial producers and consumers are directly connected to the main (380 kV, 220 kV or 150 kV) transmission grid, while medium and small producers and consumers such as greenhouse farmers and households typically are connected to the distribution grid (at 50 kV and lower).

The translations of these actors into *agents* are listed in Table 7.1. Electricity producers are agents that own and operate power plants. They sell electricity through the electricity market and buy CO_2 credits on the CO_2 market. Those markets do not own or operate any physical assets. The government is an agent which implements policies and enforces regulation, but it does not own or operate any physical assets that are part of the electricity infrastructure. Consumers are represented as households with typical demand profiles per house occupied, which combined aggregates to a typical "consumer demand".

Using and further developing the system boundary and system identification, we elected to include in this model two agents that in reality are aggregates of many individual actors: markets. By aggregating details of real markets we created suitable representations of both a physical and a non-physical market. Just as in reality, traders in the model meet in these markets to negotiate their contracts. *Markets* are considered true institutions with specific characteristics visible in the real world market arena. The power exchange is an example of such an institutionalised market; it can be seen as a power pool in which all producers are *obliged* to offer the

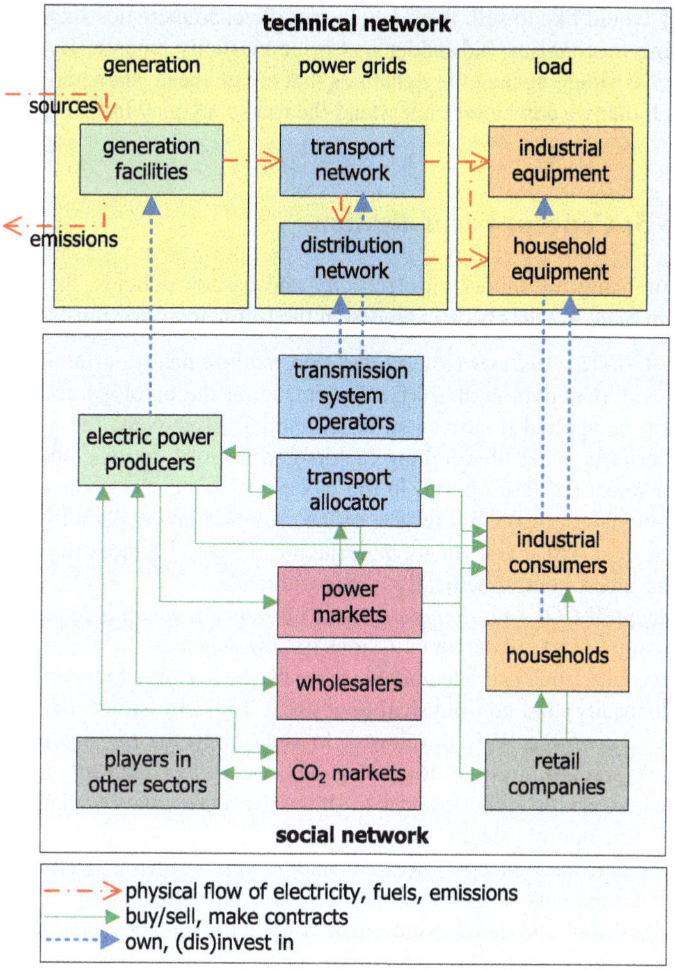

Fig. 7.1 Socio-technical systems perspective on power generation and emissions policy

Table 7.1 Agents, technologies and their relationships in the emission policies model

Agent	Relationship	Physical node
Electricity producers	owns	Power plants
Households	owns	Houses
World market	owns	Fuel equipment
Environment	owns	Environment
CO_2 market	–	–
Electricity market	–	–
Government	–	–

power they would like to sell, from which specific characteristics such as an aggregate clearing mechanism, bid ladder and price elasticity emerge. By representing the markets as single agents, the dynamics that create these characteristics are simplified, such that we can better understand the roles they play in real-life situations.

7.4 Step 3: Concept Formalisation

While formalising the model on electricity and carbon policies, the ontology, as described in Sect. 4.2, has been extended in the following directions:

- A *Multi-Criteria Analysis* (MCA) decision method has been implemented in a generic way (Chappin et al. 2007). By extending the ontology accordingly this MCA can be applied in any agent-based model. Effectively, the agents can use a set of criteria, a set of weighing factors and a set of options, and then use an MCA to select the best option. In the electricity and carbon policies model the list of options refers to the selection of the preferred power plant technology that the agent then will select for its investment. In Sect. 7.5 more details are given regarding investment, specifically in Algorithm 7.1.
- The concept of *CO_2 emission credits* was introduced, which is required to model the emission of CO_2 under an emissions trading scheme.
- To be able to include and adequately represent the trading of physical goods (e.g. fuels, electricity) and non-physical goods (e.g. CO_2 emissions credits), a variety of types of *contracts* were developed. In addition to the *PhysicalFlowContract* that allows trading physical flows of goods, the *ObjectContract* has been introduced, which is more generic and also allows the representation and formalisation of the trading non-physical goods.
- In this model, the concepts of *Agent* and *Technology* have been developed to represent completely separate entities. Thus we can formalise and represent technology portfolios and their development. The model includes individual agents—power companies—which own and strategically evaluate their entire set of technologies owned (i.e. power plants). The concept of *Ownership* was incorporated in the ontology to allow for such representation.

7.5 Step 4: Model Formalisation

In general, the formalisation described in Sect. 7.4 was chosen to allow the separation of layers of agent behaviour. Figure 7.2 displays the conceptual model of agent decisions. We discern between *strategic decisions*, which typically are about investment or the change of ownership of physical assets, *operational decisions*, which concern the day-to-day technical management and control of physical assets, and tactical decisions, which concern the deployment of assets in relation to the market. In the model, there are three main steps of agent behaviour that follows from those three layers. These are summarised in the model narrative.

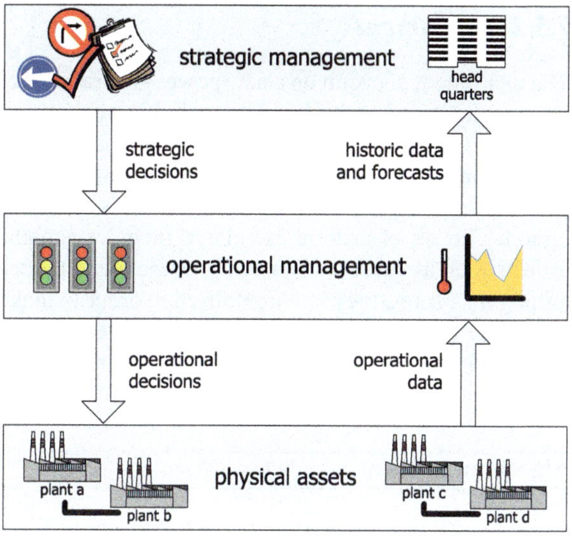

Fig. 7.2 Formalisation of agent decision-making in three layers (from Chappin et al. 2007)

7.5.1 *Model Narrative*

Upon starting the simulation which runs the electricity and carbon policies model, six electricity producer agents are populated with a set of power plants, reflecting the current status of the Dutch power generation sector. Electricity, CO_2 and fuel markets are initialised, and the government agent implements one of the emission policies. During the subsequent unfolding of the simulated power sector, electricity producers decide on investment in new plants, the dismantling of old power plants and the timing thereof. Each agent has its own personal style regarding investment. If investments are made, the related installations will only start operation after the typical lead time required for design and construction, but the investment plans are announced to the other agents immediately.

After all electricity agents have taken their (dis)investment decisions, the focus shifts to the operational aspects of power generation. The producing agents are required to bid on both the electricity and the CO_2 markets. These markets are cleared in an iterative process. Upon clearance of the markets the emissions credits are transferred and electricity sales are disclosed.

Depending on the policy implemented, the government will now do one of the following: do nothing, impose penalties upon electricity companies who do not own sufficient credits, or collect the CO_2 taxes due. The electricity companies will obtain the required amount of fuel to allow them to generate the amount of electricity contracted, and they will operate their power plants accordingly.

Below, two aspects are discussed in more detail: the investment algorithm and the market clearing algorithm.

7.5.2 *Investment*

The investment decision on a new power generation technology is taken by completing a multi-criteria analysis. Although this was formally described in another work (Chappin et al. 2007), we have transcribed the decision algorithm into pseudo-code (see Algorithm 7.1). Each agent first determines whether a need for capacity expansion exists. It then calculates scores for each relevant criterion for each alternative option. The set of criteria calculated includes economic aspects and less rational aspects such as conservativeness, greenness and perceived nuclear threat. After obtaining the scores, they are normalised in order to make them comparable. The total scores are calculated by summation. The alternative with the highest sum score is selected. However, the decision to make the investment is only taken in cases where the expected lifetime profit of the alternative selected is positive.

Algorithm 7.1 Pseudo-code for investment

```
if demandForElectricity > supplyCapacity +
    currentCapacityExpansionPlans
  listPossibleInvestmentOptions
  for possibleInvestment from
      listPossibleInvestmentOptions
  do
    for criteria from listOfCriteria
    do
      calculateScoreOnCriteria(possibleInvestment)
    end for
    normaliseScores
    multipleScoresWithWeightFactors
    sumScoreOfEachAlternative
    selectBestAlternative
  end for
end if
if expectedProfit > 0
  invest
end if
```

7.5.3 *Bidding on Markets*

The outcome of the CO_2 market is input to the power market and vice versa. Therefore, the electricity and CO_2 markets are iteratively cleared. This iteration is com-

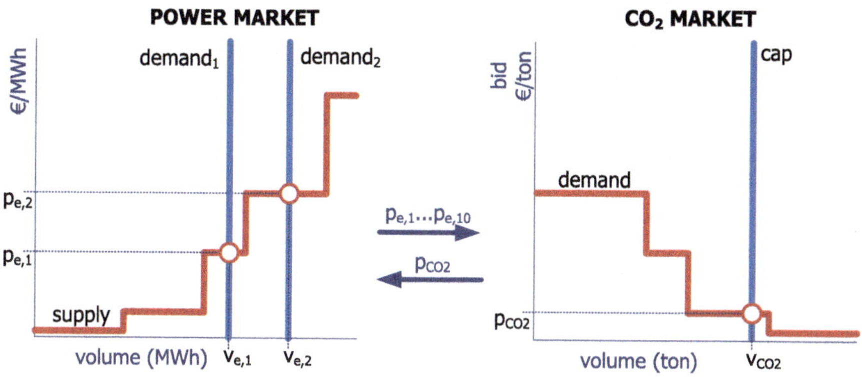

Fig. 7.3 Iterative clearing of power and CO_2 markets (from Chappin et al. 2009a)

plete when stable prices have been established for an entire year. This is depicted in Fig. 7.3 and also comprises the *while-loop* in the description in Algorithm 7.2.

Algorithm 7.2 Pseudo-code for the clearing of the markets

```
while CO2 or electricity price unstable
do
   for agent from electricityProducerAgents
   do
        agent.bidOnCO2Auction
   end for

   co2Autcion.clearTheAuction

   for agent from electricityProducerAgents
   do
        agent.bidOnElectricityMarket
   end for

   electricityMarket.clearTheMarket
end while
```

In each time step in the simulation we start with the prices of the previous year. In each iteration, first the CO_2 auction is cleared, which results in a CO_2 price (p_{CO_2}).[2]

[2]When the CO_2 price exceeds the penalty level, agents will rationally choose to pay the penalty rather than purchase more CO_2 credits. Consequently, this penalty level functions as a price cap for the CO_2 market.

This price is then used by the electricity-producing agents to calculate their power market offers. The power market is made up of ten different levels of demand (of which 2 are drawn in Fig. 7.3). These ten levels reflect the fact that demand varies over the year; they have been selected to adequately represent a so-called *load-duration curve*.

When all agents are done bidding, the bids of all the power producers are sorted and stacked. Subsequently, the market is cleared for each of the ten demand levels. This results in 10 electricity prices ($p_{e,1} \dots p_{e,10}$). In the next iteration where bids are made in the CO_2 auction, these clearing prices for electricity are used as the estimated expected price of electricity. Upon completion of this iteration, emissions trading effectively has been completed. It is not guaranteed that an equilibrium will be reached, especially because there are hard constraints that must be respected (e.g. the penalty level) and because all bids concern a relatively large volume of electricity or emission rights. Thus it is possible that the procedure results in sets of price levels on the power and CO_2 markets that hibernate between iterations. If this situation is detected, one of the states is selected at random.

7.6 Step 5: Software Implementation

The models have been implemented using a variety of software tools, of which the most important are mentioned in Fig. 7.4. The software packages have been bundled and connected into a so-called *software stack*. We have used the software stack described in Chap. 3 and developed it to accommodate our specific needs.

The software stack is used two facilitate two different tasks, the first of which is model development. The use of the programmes Protégé, Eclipse and Repast allows for completing single test simulation runs while developing the model. The second task is to create model results and explore the parameter- and scenario-space of the model by completing hundreds or thousands of runs. Since each simulation requires quite some time to complete on a PC, these sets of simulation runs have been executed on a high performance computing facility (HPC).

Repast forms the link between the first and second task because it is used to start and control each simulation run and to record the data. Torque is used to execute and manage the simulations on the HPC. Upon completion of the runs, the simulation data generated are analysed using Matlab code (Mathworks 2010), which produces statistical tests and graphs. The data are pulled from the database maintained on the HPC head node, where it is stored.

The entire modelling process is facilitated by *Subversion* (SVN). All code, model libraries, but also all scripts, Matlab code and simulation results, are stored on our SVN server. Local copies are available both on user stations (PC, workstation) and on the HPC. Changes are retrieved from and committed to the svn server. In combination with the scripts this simplifies many of the tasks surrounding the performance of simulations. The end result is a dramatic increase in productivity of the modeller who uses his PC and the HPC to complete both tasks: modelling and simulation.

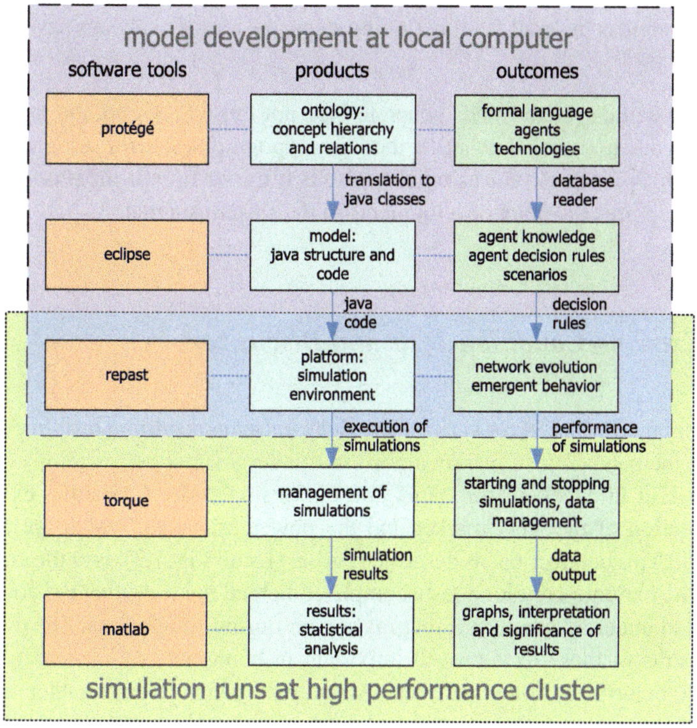

Fig. 7.4 Software model development and running and analysing simulations

7.7 Step 6: Model Verification

The verification step normally is conducted prior to the execution of sets of simulation runs. The objective of verification is to check that the model is free of errors. A wide variety of tests have been performed to verify this for the electricity and carbon policies model. Below we describe some that proved important.

7.7.1 Preliminary Simulations

The process of verification starts early in the model development (it is the first task) and consists of analysing preliminary simulations. During model development the model is run over and over again. Verification starts by looking at the indicators to check whether each evolution at the current state of the model is correct. The process has been designed to accommodate this:

- Important system indicators are identified before model development starts.
- Before any programming of behaviour commences, the graphs and graphics depicting the system indicators are programmed.

- Development is in small but functioning steps; the idea is to always have a running model.

When specific system trends generated are not those expected, the modeller inspects the code to establish whether there is a typographical error, a logical error or no error. This process produces reliable results if the set of indicators is sufficiently large and captures the main developments in the system simulated.

7.7.2 Extreme-Condition Tests and Discussion

Another method we used for verification is the extreme-condition test. In particular, checking the behaviour of market algorithms under a variety of extreme conditions proved useful in checking for errors and optimising code. A peculiar example is the interaction of the CO_2 market and the power exchange, which are mutually dependent. This connection, as described earlier (recall Fig. 7.3), is at the core of the model. The extreme-condition tests completed helped to assess that the code works as expected under all conditions that may occur during simulations. The discussion of the results of these tests proved important in being able to adequately capture real-world behaviour for both markets. An extremely high price on either one of the markets, for example, led to a far higher price on the other market. Another example is to use a very high fuel price and observe that the technologies which use these fuels are priced out of the market. After performing and analysing many of such tests we were confident that there are no errors in the code.

7.7.3 Agent Behaviour Tests

At the core of the agents is the investment decision. The multi-criteria analysis was therefore tested separately. As described above, this piece of code is generic and is attached as a module. Therefore, it is also possible to test it outside the simulation; it provides a clean testing environment.

In later evolutions of a model of CO_2 policies and power generation, we conducted an extensive range of tests on the individual investment decisions of agents. Since computation of one single decision of one single agent requires little resources (in terms of CPU time), it is possible to complete millions of such computations and simulate the investment decision under all possible relevant circumstances. Thus the most crucial aspect of the model—agent behaviour through its investment decision—can be tested. In the development of the electricity and carbon policies model, the analysis was used to show whether everything worked as expected and whether there were no unwanted artefacts in the model.

7.7.4 Repetitions

A number of replications is necessary because the behaviour exhibited by the model is different in every run. The outcomes between runs differ even when the parameters and scenarios are the same. There are a number of sources for differences in outcomes between runs:

- the order in which agents are given the opportunity to take decisions (e.g. which agent is the first to invest),
- the scenarios (e.g. fuel price developments vary between runs), and
- the behavioural rules of agents (e.g. weight factors for the multi-criteria analysis of agents vary between runs).

All these sources require repeating simulations, the exact number of which is difficult to determine. When the results stabilise, i.e. when more repetitions do not significantly affect the result, a sufficient number of repetitions is reached. In our case, this implied running a couple thousand runs.

7.8 Step 7: Experimentation

The experiments were designed to accommodate and *compare* three policy scenarios: no intervention, emissions trading and carbon taxation. Particularly the comparison between emissions trading and carbon taxation proved not to be straightforward, because under trading the market generates the CO_2 price, while under taxation this is imposed by government. The market price emerges during the simulation and is dynamic, while taxation is imposed and largely static.

We opted to make the two comparable by making sure the *average* CO_2 price over time was comparable, which translates to similar "regulatory pressure" of CO_2 pricing. This effectively means setting the taxation level to be equal to the average CO_2 market price generated in the set of simulation runs for emissions trading. In the actual EU ETS the cap will be reduced each year. We translated this to taxation by introducing the additional criterion that taxation can only remain at the same level or increase.

The electricity producers—the agents—operate in a dynamic world, which was represented as exogenous trends. These were represented as time series of fuel prices, electricity demand and carbon policy parameters (the emission cap and tax level, respectively). We assumed that the electricity producers have no market power, neither in the fuel markets nor in the electricity or CO_2 markets. This was input for the modelling of behaviour, i.e. their bidding strategy, but also for modelling the scenarios. A range of scenario values were modelled. The fuel prices in the simulation start at current market levels and develop stochastically, according to certain trends.

Since it is rather easy for the modeller to change scenario parameters according the ideas of the *model user*, the problem owner can easily repeat the analysis with a data set that corresponds to the strategic knowledge they would *not* be keen on sharing.

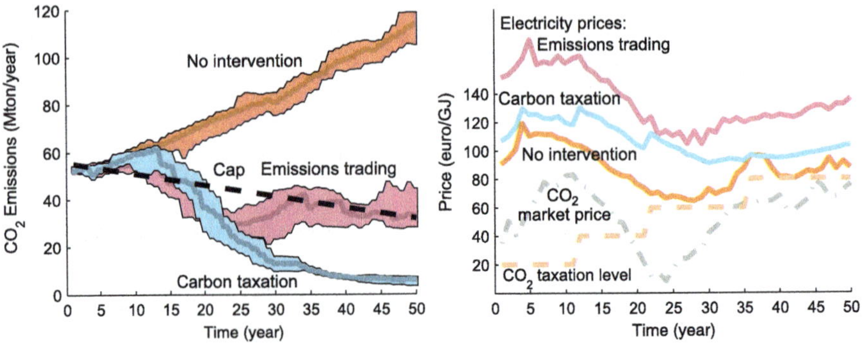

(a) *Emission levels* **(b)** *Electricity and CO_2 prices*

Fig. 7.5 Electricity prices, CO_2 prices and emission levels (adapted from Chappin et al. 2009b)

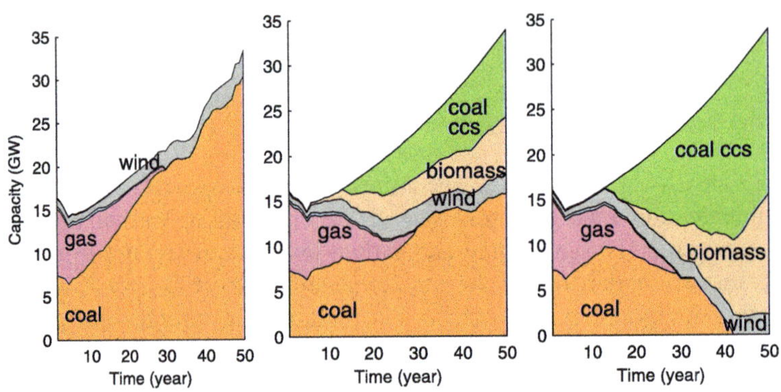

(a) *No intervention* **(b)** *Emissions trading* **(c)** *Carbon taxation*

Fig. 7.6 Average generation portfolio evolution for the three scenarios (adapted from Chappin et al. 2009b)

7.9 Step 8: Data Analysis

Since we have identified a clear list of indicators, analysing data was a straightforward task. The main indicators are system parameters over time, and we captured and depicted them in Figs. 7.5 and 7.6 as graphs over time. The evolution of the system level electricity generation portfolio is displayed using average developments under each of the policies. We display bands of the values found in the data, providing information regarding the variance in the data.

CO_2 emission levels over time is an indicator for the performance of the system and the emission policies under study. As can be seen in Fig. 7.5, with no intervention the emissions steadily increase. Under emissions trading, the emissions follow the EU regulated cap, while under carbon taxation, emission levels drop below the cap.

The prices that emerge on the electricity market and the CO_2 market are shown in Fig. 7.5. These are endogenous prices which emerge in the simulation from the interaction between the agents. The electricity price is an indicator for the cost for consumers and can help to establish the financial burden that an emissions policy will cause to consumers. It can be seen that the electricity price under emissions trading is higher throughout the simulations than that of carbon taxation. As expected, electricity prices remain lowest under a scenario of no intervention.

The CO_2 price is an indicator for the strength of the incentive of the emissions policy. The price on the CO_2 market is on average 40 euros/ton, but is very volatile: it varies between 10 and 80 euros/ton. We have implemented a taxation level that is equal on average (40 euros/ton) and which rises gradually from 20 to 80 euros/ton.

The evolution of the portfolio of power plants over time, on the system level and on the level of individual agents, show the cause for the emission levels and give insight in the investment patterns. In Fig. 7.6, the portfolio developments are depicted over time for each scenario. Coal is dominant under no intervention. Under emissions trading, coal is also important, but a more diverse portfolio emerges. Under carbon taxation, coal and gas are replaced by wind, biomass and coal used with Carbon Capture and Storage (CCS) technology.

7.10 Step 9: Model Validation

The main task in validation is to show that the model is *fit for its purpose*. As we are exploring transitions in the energy domain, specifically the transition to a low-carbon electricity supply, the question is whether we have created a model that helps us to adequately address this question. We do not claim that the model allows us to accurately predict such a transition or the actions taken by companies, governments or consumers. Yet the model does help us to identify patterns emerging from actor interaction and decisions shaping the technical system. Thus, the model offers us novel insights in the dynamics of this sociotechnical systems in transition. As discussed in Chap. 3, we cannot perform experiments in the real world to test validity. The in-silico experiments, however, were conducted after tests for validity had been completed. These include two main activities: (1) to provide a transparent model developed in an open process, and (2) to use expert opinions to validate the model.

Together, these provide ample opportunity to gain expert insight in the validity of the model. A range of experts helped in the development process. The series of models and the results generated and insights obtained have been presented and discussed on many occasions, where opportunity was given for feedback and suggestions.

7.11 Step 10: Model Use

The model was used to arrive at robust conclusions regarding the advantages and drawbacks of emission policies. The development and *use* of this model therefore has become an ongoing process spanning quite some time, during which refinements and extensions have been incorporated. During the use phase of the model, additional input was obtained from expert discussions, presentations to various stakeholders and their experts, as well as in a "serious game".[3]

7.11.1 Emergent Insights from Iterations and Discussions

The main insights emerged from the many iterations in the model development. Many iterations of the model were analysed and discussed internally by looking at the model code, refining the model and analysing graphs of the indicators. The analysis phase therefore took a relatively long period of time, in which internal debates proved fundamental for the improvement of the model and our understanding of what were useful modelling choices. In that period of time, a whole series of models was produced, ranging from the model of a very *open* emissions trading scheme in which the relationship between the electricity market and the CO_2 market is unidirectional (Chappin and Dijkema 2009), to a more closed emissions trading scheme simulating a bidirectional interaction between those markets and comparing it with a carbon taxation scheme (Chappin et al. 2009b).

Presentations were held in a variety of scientific conferences and workshops as well as non-scientific settings (e.g. NGInfra Academy,[4] the Ministry of Economic Affairs and the energy company Eneco). In some of these, the presentation was combined with a serious game. On these occasions, the simulation results assisted in the discussion on energy transition. Participants accepted the fact that the simulations are not intended for perfect prediction but do lead to insight and understanding of the transition to a low carbon electricity supply. One of the main innovative insights from the model is the inherent benefit of a carbon tax over an emissions trading scheme regarding the investment risk for power producers. The volatility of the CO_2 price in the emissions trading scheme leads to larger income transfer from consumers to producers and will probably cause less abatement than a comparable taxation scheme. This has proven to be an eye-opener in many discussions.

7.11.2 Serious Game

As has been experienced in many an educational settings, it is very difficult to convey the fundamental properties of complex systems. One option to get around this

[3] A serious game is a game which has not been designed with entertainment as its main purpose, but is intended to be used for training, education, problem solving and other "serious" activities.

[4] See http://www.nextgenerationinfrastructures.eu/academy.

is to use a serious game. We have developed such a game[5] of the electricity sector in which participants play the role of power companies. These companies must sell their electricity in the market and have to invest in new capacity to meet demand (de Vries et al. 2009). The players must take their decisions facing a variety of fundamental uncertainties, such as future fuel prices and electricity demand. In a later stage, we also implemented the entering into force of energy policy changes, of which the most important is an emissions trading scheme (de Vries and Chappin 2010).

Using the game makes participants more receptive to the fundamental complexity of the socio-technical electricity system. Participants even ask questions such as '*What would happen if we had a different scenario of natural gas prices*?' and '*Could you do a lot of repetitions of the game with automated players*?' The model is then used to answer those questions and help the participants zoom out to real-world implications of their experiences in the game. The experience from playing the game has led us to conclude that it makes it possible to get the insights from the model across to them in a deeper and more applied manner.

7.12 Conclusions

The need for a *transition* to a low-carbon society is clear. Whether such a transition can be managed or invoked by policy makers—and how—remains unclear, however. Deeper understanding of socio-technical energy system is required to be able to design policies that prove to be successful. Ex ante analysis and assessment of policies requires simulation models. At the end of the day, we intend to support the policy maker to take well-informed and robust decisions regarding energy and climate policy. The model has already proven to be insightful and has gained significant attention in the scientific world as well as from the Dutch Ministry of Economic Affairs, which is responsible for climate and energy policy in the Netherlands.

An agent-based simulation model has been developed to explore the potential of transition in our electricity supply that may prelude the departure from our fossil-based power generation infrastructure. Carbon taxation and emissions trading are policy instruments for achieving significant CO_2 emissions reduction by inducing a shift in technology and fuel choice. Our simulations of a competitive electricity generation sector show that under both policies, CO_2 emissions increase for 10–15 years due to the long life cycle of power plants. Dramatic reductions materialise after 20–40 years, when a tight cap or sufficient tax level is maintained. When taxes are set equivalent to trading prices, taxation induces earlier investment in CO_2 abatement, a better balance between capital and operating costs, and lower long-run electricity prices. In this way the simulation model has provided insight in some fundamental differences between possible emission policies.

[5] Available online at http://emg.tudelft.nl.

References

CEC (2003). Directive 2003/87/ec of the European Parliament and of the Council of 13 October 2003 establishing a scheme for greenhouse gas emission allowance trading within the community and amending council directive 96/61/ec. *Official Journal of the European Union, 275*, 32–46.

CEC (2009). Directive 2009/29/ec of the European Parliament and of the Council of 23 April 2009, amending directive 2003/87/ec so as to improve and extend the greenhouse gas emission allowance trading scheme of the community. *Official Journal of the European Union, L140*, 63–87.

Chappin, E. J. L. (2006). *Carbon dioxide emission trade impact on power generation portfolio, agent-based modelling to elucidate influences of emission trading on investments in Dutch electricity generation*. Delft: Delft University of Technology.

Chappin, E. J. L. (2011). *Simulating energy transitions*. PhD thesis, Delft University of Technology, Delft, the Netherlands. ISBN: 978-90-79787-30-2.

Chappin, E. J. L., & Dijkema, G. P. J. (2009). On the impact of CO2 emission-trading on power generation emissions. *Technological Forecasting & Social Change*, *76*(3), 358–370.

Chappin, E. J. L., & Dijkema, G. P. J. (2010). Agent-based modeling of energy infrastructure transitions. *International Journal of Critical Infrastructures*, *6*(2), 106–130.

Chappin, E. J. L., Dijkema, G. P. J., van Dam, K. H., & Lukszo, Z. (2007). Modeling strategic and operational decision-making—an agent-based model of electricity producers. In J. Sklenar, A. Tanguy, C. Bertelle, & G. Fortino (Eds.), *The 2007 European simulation and modelling conference*, Eurosis, St. Julians, Malta.

Chappin, E. J. L., Dijkema, G. P. J., & Vries, L. J. D. (2009a). Agent-based simulation of carbon policies and power generation. In *32st IAEE international conference, energy, economy, environment: the global view*, San Fransisco: IAEE.

Chappin, E. J. L., Dijkema, G. P. J., & Vries, L. J. D. (2009b). Carbon policies: do they deliver in the long run? In P. Sioshansi (Ed.), *Global energy policy and economic series. Carbon constrained: future of electricity* (pp. 31–56). Amsterdam: Elsevier.

de Vries, L. J., & Chappin, E. J. L. (2010). Power play: simulating the interrelations between an electricity market and a CO2 market in an on-line game. In *33st IAEE international conference, the future of energy: global challenges, diverse solutions*, InterContintental Rio Hotel, Rio de Janeiro: IAEE.

de Vries, L. J., Subramahnian, E., & Chappin, E. J. L. (2009). Power games: using an electricity market simulation game to convey research results. In *Proceedings of the second international conference on infrastructure systems 2009 (INFRA 2009): developing 21st century infrastructure networks*, Chennai, India.

Dijkema, G., & Basson, L. (2009). Complexity and industrial ecology: foundations for a transformation from analysis to action. *Journal of Industrial Ecology*, *13*, 157–164.

Egenhofer, C. (2003). The compatibility of the Kyoto mechanisms with traditional environmental instruments. In C. Carraro & C. Egenhofer (Eds.), *Firms, governments and climate policy: incentive-based policies for long-term climate change*, Cheltenham: Edward Elgar.

Ehrhart, K. M., Hoppe, C., Schleich, J., & Seifert, S. (2003). Strategic aspects of CO2-emissions trading: theoretical concepts and empirical findings. *Energy and Environment*, *14*(5), 579–598.

EnergieNed (2006). Energiened. http://www.energiened.nl/.

Escalante, D. I. (2010). *The EU ETS and the electricity sector: reducing emissions under market and policy uncertainty*. Master's thesis, Delft University of Technology, Faculty of Technology, Policy and Management.

Grubb, M., & Newberry, D. (2007). Pricing carbon for electricity generation: national and international dimensions. Faculty of Economics, University of Cambridge.

Hardin, G. (1968). The tragedy of the commons. *Science*, *1968*(162), 1243–1248.

Hoffmann, V. H. (2007). EU ETS and investment decisions: the case of the German electricity industry. *European Management Journal*, *25*(6), 464–474. Business, Climate Change and Emissions Trading.

IPCC (2007). *Climate change 2007: mitigation of climate change summary for policymakers.* Geneva: IPCC.

Mathworks (2010). Matlab. http://www.mathworks.nl/.

McCormick, J. (2001). *The European series: Environmental policy in the European Union.* Basingstoke: Palgrave Macmillan.

Smith, A. (1776). *An inquiry into the nature and causes of the wealth of nations.* London: Random House.

Svendsen, G. T. (1999). The idea of global CO_2 trade. *European Environment*, *9*(6), 232–237.

Svendsen, G. T., & Vesterdal, M. (2003). How to design greenhouse gas trading in the EU? *Energy Policy*, *31*(14), 1531–1539.

UNFCCC (1998). Kyoto protocol to the united nations framework convention on climate change. http://unfccc.int/resource/docs/convkp/kpeng.pdf.

Chapter 8
An Agent-Based Model of a Mobile Phone Production, Consumption and Recycling Network

L.A. Bollinger, C.B. Davis, and I. Nikolic

Abstract This chapter introduces an agent-based model of material flow networks associated with mobile phones, including production, consumption and recycling. Against a background of impending metal scarcity and growing stocks of metal waste, the case study explores the development of "closed-loop" flow systems for metals in mobile phones. Agents in the developed model are designed to represent actors in the mobile phone life-cycle, including manufacturers, consumers, refurbishers and recyclers. Via interaction with one another, these agents trade metals, mobile phones and mobile phone components, ultimately giving rise to a network of transactions. The structure of this network is an emergent property of agent decision-making and of environmental variables such as metal prices and demand growth. The aim of the modelling exercise is to gain insight into how adjustments in agent decision-making and in these environmental variables can lead to greater cyclicality of metal flows at the emergent system level. The model introduces several innovations relative to the previous cases described in this book. The first of these innovations is the implementation of goods as discrete entities with unique properties, rather than continuous flows of uniform quality. This enables agents to do detailed analyses of individual goods—in this case mobile phones—in the process of taking decisions about purchasing, processing and investments. A second novel feature of the model lies in the relationship between the model code and the ontology. Unlike previous cases, communication between these two components takes the form of a continuous dialogue rather than being a one-off occurrence at initialisation. This is essential to enabling the representation of mobile phones as discrete entities and also allows for richer data analysis and simplified verification procedures.

8.1 Introduction

From the steel in automobile frames and the copper in power lines to the tantalum in mobile phones and the neodymium in the permanent magnets of wind turbines,

L.A. Bollinger (✉)
http://wiki.tudelft.nl/bin/view/Main/AndrewBollinger
e-mail: L.A.Bollinger@tudelft.nl

K.H. van Dam et al. (eds.), *Agent-Based Modelling of Socio-Technical Systems*, Agent-Based Social Systems 9,
DOI 10.1007/978-94-007-4933-7_8, © Springer Science+Business Media Dordrecht 2013

metals are essential ingredients in a multitude of goods that fulfil human needs on a daily basis. Despite their vital importance, the continuing availability of some types of metals is questionable. Gordon et al. (2006) suggest that 'widespread adoption of certain new technologies can be expected to encounter natural limitations' in terms of metals availability, especially in cases in which metals provide a unique service. The situation becomes further exacerbated when one considers the rising metal needs associated with growing levels of consumption in the developing world and the increasing global population. It is estimated that fulfilling the consumption needs of a global population with current technologies and consumption patterns comparable to those in the developed world would require the lithosphere's entire reserves of copper and zinc ore, and likely that of platinum as well (Gordon et al. 2006).

The relative scarcity of some metals, combined with their crucial role in fulfilling human needs, has led to the classification of certain metals as *critical*. Metals often classified as critical include platinum group metals, rare earth metals, indium, manganese, niobium, lithium, tantalum, gallium, tellurium and several others. In today's economy, such metals are most commonly applied in small amounts in various high-tech products like consumer electronics, fuel cells, batteries and solar panels. A number of them are also essential components in certain renewable energy technologies, making their continued availability crucial to the realisation of sustainable energy systems.

To address this looming issue of metal scarcity, the field of industrial ecology proposes closing metal flow loops, i.e. diverting metals in end-use products from disposal back to productive use (Ehrenfeld and Gertler 1997; Frosch and Gallopoulos 1989; Lowe and Evans 1995). Realising this in practice, however, is a challenging task. Metals are often applied in products with highly complex and intransparent life-cycles. The structure of metal flows in such systems is determined not by any single actor but by the actions of a large number of autonomous decision-makers. A good example of this is mobile phones, which contain high concentrations of precious and critical metals such as gold, palladium, indium and tantalum. While industrial-scale recycling processes are capable of recovering the vast majority of these metals (Rochat et al. 2007), most used mobile phones disappear into the drawers of consumers or tumble down the global economic ladder to countries where efficient recycling practices are far from the norm (Mooallem 2008). Exacerbating this situation is the high replacement rate of mobile phones—although their technical lifetime is on the order of 10 years, consumers normally replace their mobile phones after only 1 to 2 years (Geyer and Blass 2009).

Measures to improve this situation have thus far produced only limited results. The problem is that these measures often do not adequately account for the complexity of the system. The mobile phone product system is composed of a large number of autonomous decision-makers, from multinational original equipment manufacturers (OEMs) to small- and medium-sized refurbishing companies and individual consumers. These actors are embedded in a dynamic environment of fluctuating metal prices, explosive demand growth and rapidly evolving technological capability. Their interactions with one another and with this dynamic context gives rise to a highly complex and constantly evolving web of feedback processes.

The sections below follow the format outlined in Chap. 3. Following a summary of the problem being addressed by the model and the actors involved, the processes of system decomposition and concept formalisation are discussed. A particular challenge in concept formulation in this case was the translation of the "closed-loop" concept into a set of usable metrics. Next, the formalisation of the model is described in the form of a narrative complemented with pseudo-code, and the software implementation is elaborated.

After a brief explanation of the applied verification procedures, the experimentation process is described. This process involved the execution of 3 parameter sweeps and the exploration of a 15-dimensional parameter space. Finally, the results from analysis of these sweeps is summarised and their worth to the involved stakeholders is explained. In the conclusion to this chapter, the contributions of this model to the metals domain and to the further development of agent-based modelling of socio-technical systems are discussed.

8.2 Step 1: Problem Formulation and Actor Identification

The central problem being addressed by this model is the low recovery rate of critical metals in mobile phones. The problem owner in this research was a large recycling company which focuses on the recovery of precious and other non-ferrous metals from complex waste streams. This company recycles large amounts of electronics waste—including mobile phones—and thus has an interest in understanding the dynamics of mobile phone flow systems and how higher metal recovery rates can be realised in practice. Because the problem of metal scarcity is one of broad societal consequence, decisions taken to address it also affect a much wider array of stakeholders. The problem owner and an academic expert in mobile phone recycling and reuse were involved as domain experts during the course of the project. They provided key data with regard to the characteristics of technologies (e.g. metal recovery rates of recycling processes) and assisted in the formulation of decision-making behaviour for the defined agents. These experts provided feedback at multiple points during the development of the model and played an essential role in the validation process.

Within this context, the aim of the modelling exercise was to gain insight into the dynamics of global mobile phone flows and to develop an understanding of how micro-level adjustments—modifications in the behaviour of actors and in the properties of physical goods in the system—can lead to greater cyclicality of metal flows at the emergent system level. The research question that guided the model development was the following: *What conditions foster the development of a closed-loop flow network for metals in mobile phones?*

8.3 Step 2: System Identification and Decomposition

As described in Chap. 3, system identification and decomposition is a social process. In the case of the model described here, this process involved repeated consultation

Table 8.1 List of agent types in the model and the technologies they possess

Agent type	Relationship	Technologies
Manufacturer A (manufacturer of low-end, low metal-content phones)	owns	Manufacturing technology A
Manufacturer B (manufacturer of low-end, high metal-content phones)	owns	Manufacturing technology B
Manufacturer C (manufacturer of high-end, low metal-content phones)	owns	Manufacturing technology C
Manufacturer D (manufacturer of high-end, high metal-content phones)	owns	Manufacturing technology D
New phone retailer	owns	Retail technology
New phone retailer	owns	Collection technology
New phone retailer with collection incentives	owns	Retail technology
New phone retailer with collection incentives	owns	Collection technology
Used phone retailer	owns	Retail technology
New phone consumer	owns	Consumption technology
New high-end phone consumer	owns	Consumption technology
New or used phone consumer	owns	Consumption technology
New or used high-end phone consumer	owns	Consumption technology
Used phone consumer	owns	Consumption technology
Phone collector	owns	Collection technology
Refurbisher	owns	Testing technology
Refurbisher	owns	Refurbishing technology
Refurbisher	owns	Disassembly technology
Industrial recycler	owns	Industrial metal recovery technology
Backyard recycler	owns	Backyard metal recovery technology

with both the problem owner and an academic expert in mobile phone recycling and reuse. These discussions helped us to identify and structure the elements and relationships essential for representing the system with respect to the research question.

The modelled system is composed of the actors, processes and goods relevant to global flows of metals in mobile phones. The model consists of a set of 16 different types of agents, defined to represent the different types of actors in this system. These agents and their associated technologies (physical nodes) are listed in Table 8.1. As this table indicates, each agent type has a characteristic technology or set of technologies, and every agent of that type will possess the same technology or set of technologies.

As with several of the models presented in previous chapters (e.g. Chap. 7), the external world is represented by a world market agent and an environment agent. In the case of the model described here, the world market agent represents the global metals market, as well as the markets for used mobile phone components and spent

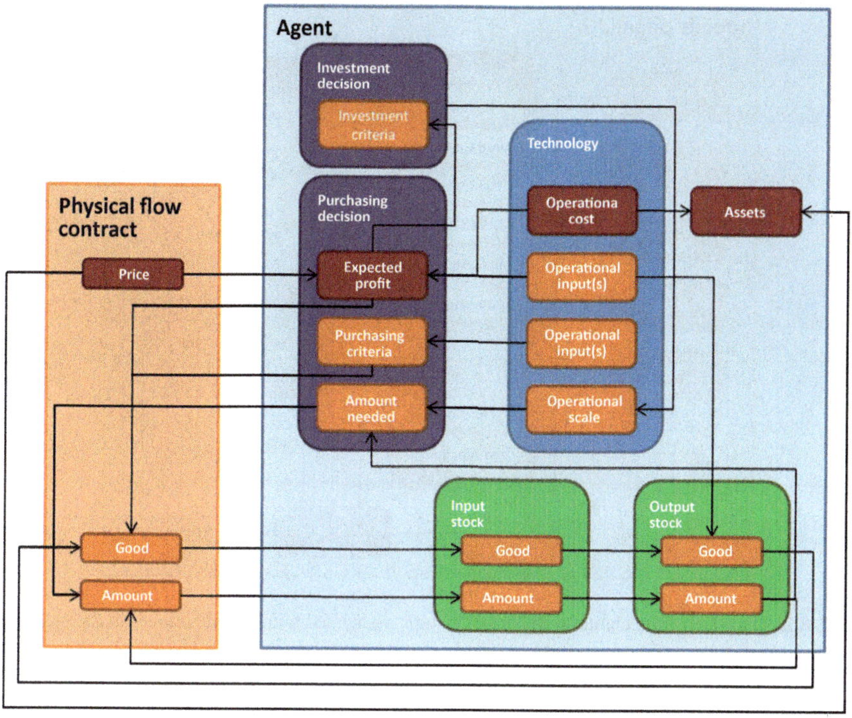

Fig. 8.1 Diagram of the elements and relationships associated with agent states and behaviours in the model

batteries. The environment agent is a disposal sink, collecting mobile phones disposed by consumers and metals lost by recycling processes.

The structure of an agent in the model is illustrated in Fig. 8.1. Agents make two types of decisions: *purchasing decisions* and *investment decisions*. As illustrated in Fig. 8.1, these decisions are a function of certain variables, such as the expected profit and the amount of a good needed by an agent. Each agent also possesses a particular *technology*, such as a manufacturing technology or a recycling technology, that enables it to process purchased goods in a certain way. When an agent purchases a good, it places this good in the agent's *input stock*. After it has processed that good, it places it in the agent's *output stock* and offers it for sale. Agents interact with one another by way of *transaction contracts*.

Goods in the model include mobile phones, batteries, mobile phone components and four different types of metals: gold, copper, silver and palladium. As illustrated in Fig. 8.2, mobile phones are described by a set of properties (age, technical lifetime, condition, functionality, etc.). The value or state of these properties is unique to each mobile phone *instance* in the ontology and changes over the course of a simulation. In taking purchasing decisions and in processing mobile phones, agents are able to view and adjust certain of these properties in different ways.

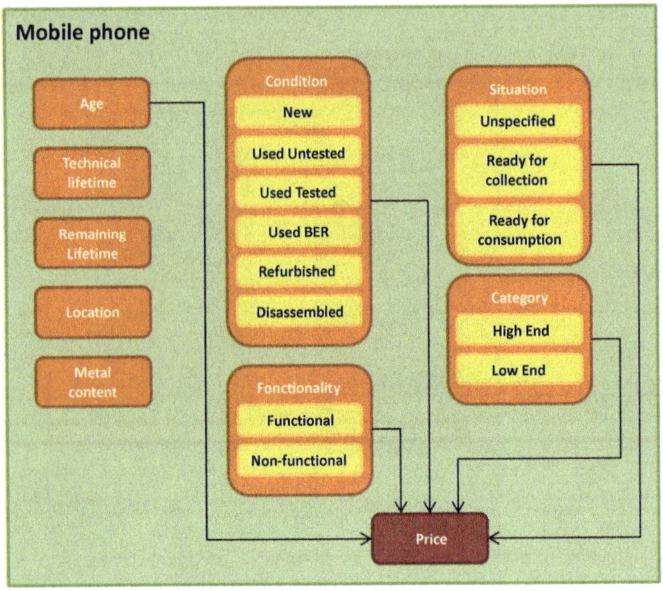

Fig. 8.2 Diagram of the elements and relationships associated with mobile phones in the model

8.4 Step 3: Concept Formalisation

The process of formalising the concepts identified and structured during system decomposition involved modifying and adding classes and instances to the ontology introduced in Sect. 4.2. The most significant of these changes was the addition of the mobile phone as an ontological class and the definition of a set of mobile phone properties (illustrated in Fig. 8.2). These properties (e.g. condition, functionality or age) represent mobile phone features relevant to the purchasing and processing decisions of various agents. Several of these properties have a limited number of discrete states, described as strings. For instance, the functionality of a mobile phone can only be *functional* or *nonfunctional*, and the category can be *high-end* or *low-end*. These discrete states are the instances of the functionality and category classes. Other mobile phone properties, such as age and remaining lifetime, are described numerically and have a (theoretically) infinite set of possible states. Table 8.2 summarises the properties and possible states of mobile phones as implemented in the model.

Metals in the model are represented as portions of a continuous flow, meaning that all portions of a metal flow between agents have the same properties. This manner of representing goods, however, was determined to be unsuitable for mobile phones since the properties of individual phones may play a large role in determining how they are dealt with in the real world. For this reason, mobile phones are represented in the model as discrete entities—independent objects that can be viewed and modified individually by agents in the modelled system.

Table 8.2 Properties and possible states of mobile phones

Property	Possible states
Condition	new, used untested, used tested, refurbished, beyond economic repair
Category	high-end, low-end
Functionality	functional, non-functional
Situation	unspecified, ready for consumption, ready for collection
Age	an integer value
Remaining lifetime	an integer value
Gold content	a floating numeric value
Copper content	a floating numeric value
Silver content	a floating numeric value
Palladium content	a floating numeric value

The representation of mobile phones as discrete entities was technically enabled by manifesting each mobile phone in the model as a unique instance in an RDF data store (see Chap. 9). Each time a mobile phone is "manufactured" during a simulation, an instance of that phone is immediately added to the data store. As the simulation proceeds and more phones are manufactured, additional phone instances are added to the data store. Furthermore, each time the properties of a mobile phone are modified by an agent, this is reflected in the phone's description in the data store.

In addition to modifying and adding classes and instances to the ontology, the concept formalisation step also involved formalising the concept of a closed-loop flow network. For this purpose, we developed a set of two metrics that could be applied to gauge the cyclicality of metal flows in the modelled system. The first of these metrics is the *closed-loop indicator (CLI)*, which is intended as a measure of the degree to which metals are preserved within the system. The CLI is mathematically described as follows:

$$CLI_m(n) = 1 - \frac{m_{disposed\ phones}(n) + m_{disposed\ by\ recycling}(n) + m_{hibernating}(n)}{\sum_{t=0}^{n} m_{manufacture}(n)} \quad (8.1)$$

where:

$CLI_m(n)$	is the value of the CLI for metal m at time n,
$m_{manufacture}(n)$	is the mass of metal m used in the manufacture of mobile phones at time n,
$m_{disposed\ phones}(n)$	is the mass of metal m in the stock of disposed phones at time n,
$m_{disposed\ by\ recycling}(n)$	is the mass of metal m in the stock of metal disposed by recycling at time n,

$m_{hibernating}(n)$ is the mass of metal m in the stock of hibernating[1] phones at time n.

The CLI reveals the fraction of metals that have either been recovered or are circulating dynamically through the system. Its value can vary between 0 and 1, with a value of 1 indicating ideal closed loop behaviour (i.e. no metals disposed to the environment or immobilised in the hibernating stocks of consumers), and a value of 0 indicating that all metals have been either disposed or immobilised.

The second metric defined is the mass of gold disposed, assuming all phones in hibernation at the end of a simulation are recycled. The value of this metric is calculated as follows:

$$m_{total\ disposed}(n) = m_{disposed\ phones}(n) + m_{disposed\ by\ recycling}(n) + m_{hibernating}(n) * (1 - r) \quad (8.2)$$

where r is the fraction of metals recovered by metal recovery processes.

The use of both of these metrics, rather than the CLI alone, reflects an assumption that closed-loop performance has not only to do with the fraction of metals preserved in the system but also with the absolute amounts of metals disposed.

8.5 Step 4: Model Formalisation

The basic series of actions undertaken by agents in the model includes the creation of contracts, the purchasing of goods, the processing/manufacture of new goods and investment in new technologies. The narrative and pseudo-code below describe each of these actions.

8.5.1 Create Contracts

The first step is for each agent to create a sale contract for each of the goods in its output stock. These contracts contain all the necessary information about a good for agents to be able to take purchasing decisions. This includes basic information such as the type of good (e.g. mobile phone, gold, battery), the price of the good and the amount being offered for sale. If the good being offered is a mobile phone, the contract also contains specific information concerning the properties of the good, for instance its functionality, condition and category. For goods such as metals, batteries and components, prices are pre-defined. For mobile phones, prices are determined by a combination of values found in the literature (market price) for mobile phones of different types and rules described in pseudo-code in Algorithm 8.1.

[1] Hibernating stocks refer to end-of-use phones placed indefinitely in storage by consumers.

Algorithm 8.1 Mobile phone pricing rules

```
if situation of phone is 'ready for consumption'
                  set price to market price * 1.1
        else if situation of phone is ready for
             collection
                  set price to 0
        else if situation of phone is unspecified
                  set price to market price
end if
```

8.5.2 *Purchase*

The next step is for each agent to analyse the list of available contracts and to choose which goods from this list to purchase. To do this, each agent looks at its own output stock and unique purchasing criteria to determine how much it needs to purchase and which criteria potential purchases must fulfil. Different types of agents have different sets of criteria. For instance, a consumer of new, high-end phones looks only for phones that have condition "new", category "high-end", functionality "functional" and situation "ready for consumption". A consumer of used phones looks only for phones with condition "used tested" or "used untested", functionality "functional" and situation "ready for consumption". In cases where more goods fitting an agent's purchasing criteria are available than what that agent can purchase, it prioritises its purchases to maximise expected profit. This is described by the pseudo-code in Algorithm 8.2.

8.5.3 *Process*

The next step is for each agent to use its available technological capabilities to process a good into a different form. Each agent—including consumers—is conceptualised to possess a technology that enables it to transform a good or its properties in a particular way. For instance, a manufacturer transforms defined amounts of gold, silver, copper and palladium into mobile phones; a new phone consumer transforms the condition of mobile phones from "new" to "used untested"; and a retailer transforms the situation of mobile phones from "unspecified" to "ready for consumption". In some cases, there are multiple ways in which an agent can modify the properties of a good it has purchased. For instance, a refurbisher can choose either to refurbish a phone it has purchased, to disassemble it or simply to resell it as is. When confronted with such a choice, an agent will always choose to process a good in a manner that maximises the agent's expected profit.

Algorithm 8.2 Mobile phone purchase prioritisation rules

```
for each technology agent that is not a consumer
        calculate how many phones you want to buy
        make list of all available phones that fit buying criteria
        calculate the expected profit of each phone on this list
        sort list according to expected profit

        while you still need to buy more phones
                select phone with the highest expected profit
                sign a contract to purchase phone
                delete phone from list of available phones
        end while
end for

for each consumer agent
        calculate how many phones you want to buy
        make list of available phones that fit buying criteria
        sort the list according to price

        while you still need to buy more phones
                select cheapest phone on the list
                sign a contract to purchase phone
                delete phone from the list of available phones
        end while
end for
```

In the course of processing a good, some agents also determine what other types of agents are able to purchase a good they put up for sale. This is the case, for instance, with consumer agents, which decide what to do with their end-of-use phones as described in Algorithm 8.3.

In this manner, a highly motivated consumer agent can sell its phones to many collector agents. A less motivated consumer agent has fewer, if any, potential buyers, and is more likely to keep its phone in its output stock or dispose of it to the environment agent.

Algorithm 8.3 Mobile phone selling to collector rules

```
for all end-of-use phones
        if motivation level of consumer >=
            accessibility level of collector
                sign contract with collector
        else if motivation level of consumer <
            accessibility level of collector
                do not sign contract with collector
        end if
end for
```

8.5.4 *Invest*

Investment is carried out by a set of inactive agents "sitting on the sidelines" of the model. In each time step, these agents observe what is happening in the model and decide whether or not they want to invest in a technology and start actively purchasing, transforming and selling goods. Investment decisions are a two-step process. Each type of agent first checks whether there would be an oversupply of the outputs it would be producing if it were to invest in a new technology. If there is already an oversupply, then it does not invest. If there is not an oversupply, it checks whether it could make a profit from the excess inputs currently available in the market. This is described by the pseudo-code in Algorithm 8.4.

8.6 Step 5: Software Implementation

The model described here was written mostly in Java using the Eclipse integrated development environment. Concepts and relations, including technologies and goods, were defined in an ontology. The information in this ontology was used to populate initial data at the start of a simulation. Unlike other models described in this book, however, data produced during the course of a simulation were mostly kept in an RDF data store rather than in the form of Java objects. This allowed for the inclusion of complex agent queries in the model code and was technically enabled through the use of the Jena Semantic Web Framework and the embedding of SPARQL queries in the model code. The advantages of this and the manner in which it was implemented are described in more detail in Chap. 9.

8.7 Step 6: Model Verification

In the case of the model presented here, verification at the multiformal knowledge level has been accomplished by developing and confirming the suitability of the

Algorithm 8.4 Mobile phone investment decision rules

```
for all template technologies other than consumers
        determine if there is an oversupply of your
            outputs

        if there is an oversupply
                do not invest in a new technology of
                    your type
        else if you are a manufacturer
                invest in one new technology of your
                    type
        else
                make list of contracts not signed by
                    any agents
                calculate expected profits of item in
                    contracts
                if expected profit for any item > 0
                        invest in a new technology
                end if
        end if
end for
```

model pseudo-code together with the participating domain experts while also comparing it with the data and information collected from domain literature. Verification at the simulation level has been achieved by a series of unit tests, by a repeatability test and by a mass balance test, each of which compare expectations derived from the decomposed system with actual results.

As stated in Chap. 3, unit testing involves executing separate tests over different portions of the simulation code and confirming that their results correspond to expectations. In the case of the model presented here, the portions of code tested included those pertaining to the purchasing of goods, transformation of goods and investment in new technologies by each type of agent. Repeatability tests involved running the simulation many times with fixed parameter settings and checking how the degree to which the results varied conformed to expectations. A mass balance test was carried out by a piece of code which ensured that the mass of metals in the system—both as independent stocks and embedded in mobile phone stocks—remained constant throughout the duration of a simulation.

In the case of both the unit tests and the mass balance test, an attempt was made to evaluate not only typical parameter settings but to define the boundaries of the parameter space (extreme value testing) and determine where agents begin to perform unexpectedly or erratically. The purpose of this was not only to confirm that the model performed as expected, but also to generate an understanding of the boundaries within which the model could be expected to give reasonable outcomes and beyond which its results might no longer be credible.

8.8 Step 7: Experimentation

Experimentation with the model consisted of a set of 33,516 simulation runs over which 15 parameters were varied. The experimentation process was divided into three parts:

1. A parameter sweep (sweep 1) during which the values of each of 15 different parameters were varied in isolation, with the values of all other parameters remaining at their defaults.
2. A parameter sweep (sweep 2) during which the values of each of 8 different parameters were varied simultaneously, such that each possible value of each parameter was tested at each possible value of each other parameter.
3. A parameter sweep (sweep 3) with parameter values set by Latin Hypercube Sampling (Swiler 2004), which generates a defined number of parameter sets that are guaranteed to be evenly distributed across the parameter space.

The exact parameters tested in sweeps 1 and 2 and the ranges over which they were varied are summarised in Table 8.3. The parameters tested in the third sweep and the ranges over which they were varied are equivalent to those in the first sweep. The parameters for all sweeps were selected in discussion with the problem owners and with domain experts. They are those system elements that, it was hypothesised, might have the potential to challenge the modelled system to exhibit closed-loop performance.

For all sweeps, a minimum and maximum value—determined in consultation with domain experts—was chosen for each parameter. The values of the parameters were then varied at defined increments within these ranges. In setting the values for these increments, nothing was assumed about the behaviour of the system—all portions of the chosen range for each parameter were examined with the same resolution. In the second sweep, some of the ranges were decreased and the increments increased so as to reduce the number of simulations required.

The magnitude and resolution of the parameter space tested in the first and second sweeps precluded the possibility of carrying out multiple repetitions. This resulted in a lack of knowledge concerning the variability of system performance across the parameter space, a deficiency that was remedied by sweep 3. Sweep 3 involved running the model 100 times at each of the 100 different sets of parameter values. The purpose of this was to provide a sense of the variability in the performance of the model with respect to multiple metrics. In this manner, sweep 3 was used to gain a general sense of the degree of variability in the results across the parameter space, and hence the degree to which the results from sweeps 1 and 2 can be trusted.

Table 8.3 List of parameters

	Parameter	Explanation	Sweep 1 range	Sweep 2 range
1	Metal price fraction	Fraction by which metal prices are higher or lower than their 2006 levels	−0.5–0.5	N/A
2	Incentive cost	Monetary incentive paid to consumers by collectors	US$ 0–50	US$ 0–50
3	Accessibility of collection pathways	Accessibility of mobile phone collection pathways to consumers	0–1	N/A
4	Consumer motivation	Innate motivation level of consumers to deposit their end-of-use phones with appropriate collection pathways	0–1	0–1
5	Mobile phone use time	Amount of time consumers use a mobile phone	2–20 quarters	N/A
6	Consumer disposal tendency	Innate tendency of consumers to dispose of their end-of-use phones	0–1	N/A
7	Technical lifetime of a phone	The time after a phone is manufactured until it becomes non-functional	4–60 quarters	8–48 quarters
8	Price for "other components"	Price of mobile phone components	US$ 0–50	N/A
9	Avg. price of a new phone	Average price of a new phone for sale by manufacturers	US$ 0–300	US$ 0–180
10	Growth rate of new phone consumers	Probability of adding a new phone consumer each quarter	0–1	N/A
11	Growth rate of new high-end phone consumers	Probability of adding a consumer of new high-end phones each quarter	0–1	N/A
12	Growth rate of cheap phone consumers	Probability of adding a consumer of new or used mobile phones each quarter	0–1	0–1
13	Growth rate of cheap high-end phone consumers	Probability of adding a consumer of new or used high-end phones each quarter	0–1	0–1
14	Growth rate of used phone consumers	Probability of adding a used phone consumer each quarter	0–1	0–1
15	Fraction recycled by industrial metal recovery	Fraction of recycled phones processed by industrial metal recovery operations as opposed to backyard operations	0–1	0–1

In all phases of experimentation, a single simulation run consisted of 100 time steps, with each time step representing a quarter year of real-world time. Data relating to the characteristics of contracts, agent stocks, agent assets and phones was collected for each time step of each run. During the course of the simulations, this data was stored in text files and was later transferred to a database. In total, approximately 400 gigabytes of data were collected. The amount of data collected and the richness of information contained in this data is unique in comparison to other case studies described in this book. It facilitated the identification of patterns and the extraction of insights that would not otherwise have been possible.

8.9 Step 8: Data Analysis

Data resulting from the first, second and third parameter sweeps were analysed separately. In each case, the two metrics described above—the CLI and the amount of metal disposed—were used to gauge the behaviour of the model with respect to the aim of the investigation. These metrics were calculated only for a single metal: gold. The sections that follow highlight some results from the data analysis process. For a full overview of these results, the reader is directed to Bollinger (2010).

8.9.1 Default Case

Analysis of the model results for the default set of parameter values served to illuminate the characteristics of the model's baseline behaviour. The setup of the model allowed us to track the state of individual mobile phones and agents throughout the duration of a simulation. This is illustrated in Fig. 8.3, where we can see the evolution of individual mobile phones over the duration of a simulation run at the default parameter settings. Each whole number value on the vertical axis of this plot represents a single mobile phone in the model. Each circle or point in the plot represents the state of an individual phone at a particular time step during the simulation (approximately 1000 phones are instantiated during the course of this simulation).[2] The shade of each circle shows the agent owning the phone at that particular time. The size of the circle indicates its market price, which is a function of multiple properties of the phone at that point in time. Looking from left to right in this plot, we can see how the value and ownership of individual phones change over time during this single simulation run.

This plot illustrates several interesting phenomena about the default case. First, there is very little reuse occurring in the system, with the vast majority of phones being recycled or disposed after one use period. As indicated by the abundance of light gray "trails" on the right-hand side of the plot, a large proportion of end-of-use phones are simply disposed by consumers after a single use, and all phones are either disposed or recycled within 30 quarters of their manufacture. Second, we can see that no more phones are refurbished after about 75 quarters. Closer examination

[2]The number of phones instantiated during a simulation is not predefined but emerges as multiple factors play out during the course of a simulation.

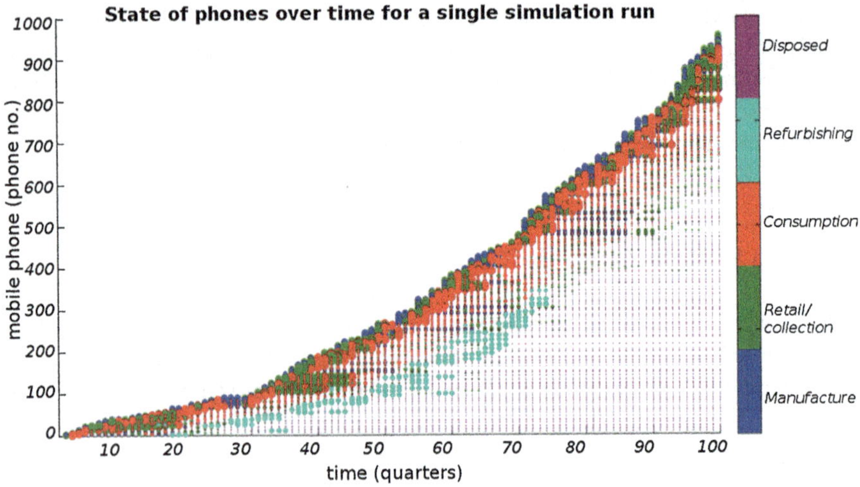

Fig. 8.3 State of phones in the system over time for the default case

of the results reveals that this is due to the "demise" of the refurbisher agent, who continually loses money from the start of the simulation and eventually goes out of business.

8.9.2 Sweep 1

The results from the first parameter sweep are summarised in Fig. 8.4, which shows the ranges of the CLI value and the mass of gold disposed produced by isolated variation of each parameter within its tested range.

The results in Fig. 8.4 indicate that the greatest leverage in terms of increasing the CLI value lies in increasing the mean consumer motivation level and the fraction of phones recycled by industrial metal recovery. In addition to producing the largest ranges of CLI values, variation of these parameters over their tested ranges also generates the highest indicator values of approximately 0.71. The lower plot in Fig. 8.4 shows the range of disposed gold values produced by variation of each parameter, assuming all phones in hibernating stocks at 100 quarters are recycled. Here, we can see that the greatest range of generated values is produced by variation of the mean use time parameter, while the lowest mass of disposed gold (2.5 g) is generated by a reduction in the average price of new phones.[3]

One notable aspect of the results from sweep 1 not adequately captured in Fig. 8.4 is the impact of increasing the mobile phone's technical lifetime. More than any

[3]While the masses of gold disposed can be compared across runs, they are not intended to accurately reflect the magnitude of real-world flows.

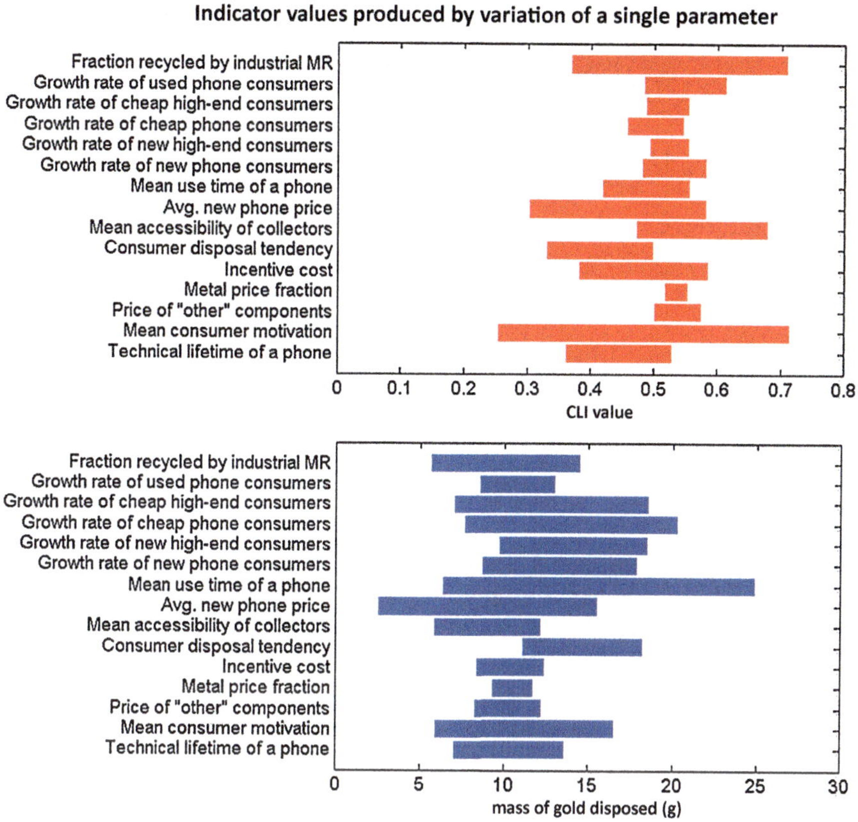

Fig. 8.4 Indicator values produced by variation of the parameter in isolation

other parameter, increasing the technical lifetime generates an increase in the fraction of phones reused. The reason for this is that higher technical lifetimes enable refurbishers to process phones more profitably. Instead of going out of business as they do in the default case, refurbishers are able to make a healthy profit. This profit enables refurbishers to remain in business and even increase their operation scales. As a result, the availability of used phones increases and certain types of consumers alter their consumption patterns accordingly. Of course, for this to occur, sufficient demand for used phones is also essential, which is a function of the growth rates in consumers of cheap and used phone consumers (see Table 8.3).

8.9.3 Sweep 2

Figure 8.5 shows the final CLI value (at 100 quarters) for each of 23,328 runs of the model at different parameter settings. As this plot indicates, it is possible within

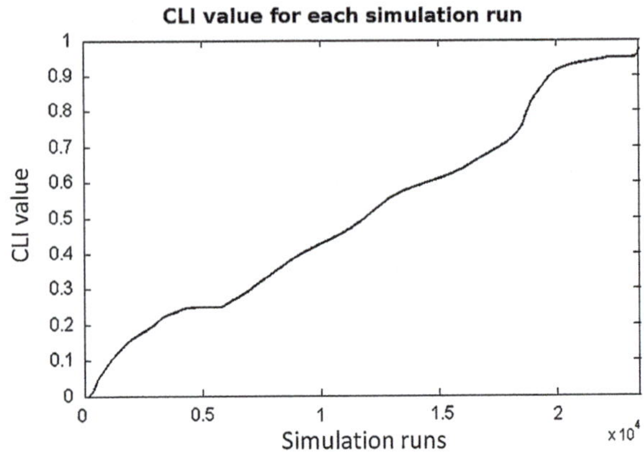

Fig. 8.5 Distribution of CLI values over all simulation runs

the ranges of parameter values tested to achieve a CLI value as low as 0.00 and as high as 0.98, with approximately an equal distribution of points above and below the 0.5 mark.

Figure 8.6 illustrates the relationship between the CLI value and the mass of disposed gold over the tested parameter space. As this plot illustrates, the mass of gold disposed varies from a low of 0.11 g to a high of about 56 g, with the majority of runs (82 %) generating less than 10 g of disposed gold. We can see here that there is a generally negative relationship between the mass of disposed gold and the CLI value, and a generally greater range of possible disposed gold values at lower CLI values than at higher ones. In this plot, we can also see that larger masses of manufactured gold generally correspond with larger masses of disposed

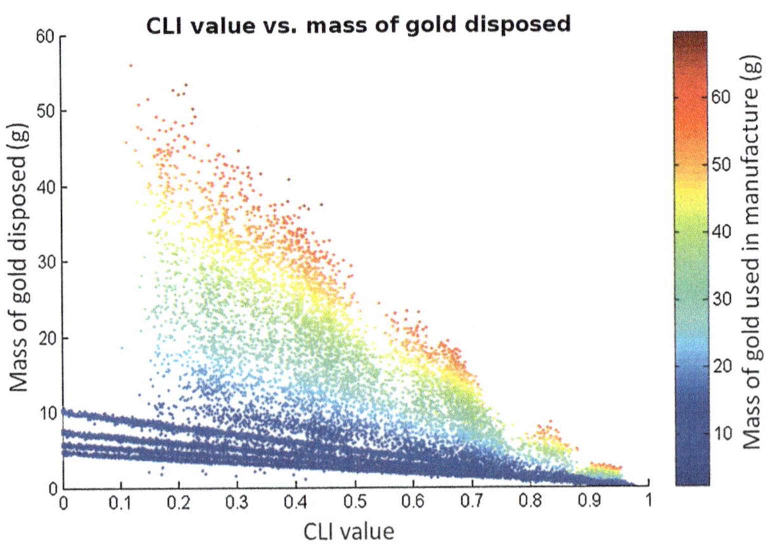

Fig. 8.6 CLI value vs. mass of gold disposed for all simulation runs

gold, indicating that a major portion of the range of disposed gold values is due to variations in the amount of gold entering the system rather the manner in which this gold is dealt with by agents within the system.

A number of further analyses were conducted on the data from sweep 2 in order to glean a better understanding of the relationship between different parameter values and the closed-loop behaviour of the system. For instance, the simulation runs that produced the top 10, 100 and 1000 CLI values were analysed to determine which parameter values were *essential* to achieving desired system behaviour and which parameter values most often co-occurred in these cases. Analyses were also performed of the impact of different parameter conditions on the financial performance of agents in the model. These analyses revealed that high rates of mobile phone reuse may be very detrimental to the financial performance of certain agents, especially manufacturers.

8.9.4 Sweep 3

Aside from looking at the standard deviation of indicator values across the parameter space and quantifying the variability of results, the results from sweep 3 allowed us to search for patterns in this variability. For instance, it was found that the value of one parameter in particular—the mean motivation level of consumers—exhibited a high degree of correlation with the standard deviation of the amount of disposed gold. This is interesting because the mean motivation level was also the parameter determined to have the most leverage over the amount of disposed gold in sweep 1.

The form of variation in the results of the model was also examined by plotting the assets of each agent over time for each of the 100 unique parameter sets tested in sweep 3. In most cases, the end states of each agent's assets were relatively evenly spaced within a particular range, or clustered around a specific value. However, in a minority of cases, bifurcations were evident, with the end state of an agent's assets clustered around two different values. An example of this is shown in Fig. 8.7. A disproportionate number of apparent bifurcations were found to have been produced by a single set of parameter values and for certain agents: the refurbisher agent and the metal recovery agents. Further investigation is required to determine whether these are indeed bifurcations, and if so why these particular combinations of agents and parameter conditions tend to give rise to such patterns.

8.10 Step 9: Model Validation

Validation of the model was carried out in two ways, both with a strong "social" component (see Chap. 3) in the sense that they involved repeated consultation with industry experts. First, the validity of the decomposed system and of the identified agent behaviour were confirmed with industry experts and compared with domain

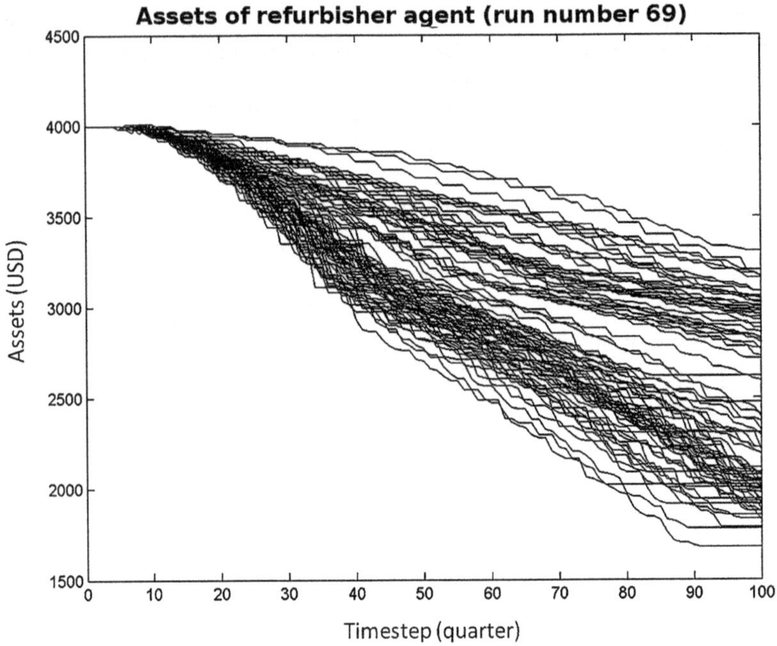

Fig. 8.7 Assets of the refurbisher agent over 100 runs at a single set of parameter values

literature. This component of the validation process was intended to ensure that the model accurately represents essential aspects of the real-world system. Second, the results of various versions of the model and the design of experiments were presented to and critiqued by industry experts. This was intended to ensure that the model in its given form is capable of providing useful insights.

Together, these two components of the validation process were meant to provide a degree of certainty that the model is both sufficiently accurate in its reflection of the real world and includes the essential ingredients to serve its intended use. The validity of the model, however, is still lacking in certain respects—for instance with respect to the decision-making rules which model the behaviour of consumer agents. Further "face validation" with experts from other domains is necessary to ensure sufficient validity of the results.

8.11 Step 10: Model Use

Multiple meetings were held with the problem owner in the course of developing the model. As described above, these meetings were an important component of the system decomposition and validation processes. However, especially as the model matured and simulation results became available, these meetings also served as a learning opportunity for the problem owner. In particular, the model results served to illustrate how various developments in the studied system and various decisions by actors in the system might affect opportunities for metal recovery. For instance, the results illustrated a strong positive correlation between the technical lifetime of mobile phones and the reuse fraction—the average number of times a phone is reused. However, the results also indicated that higher reuse fractions would significantly diminish the financial performance of manufacturers. Since manufacturers are largely responsible for determining the technical lifetime, this suggests that longer technical lifetimes may be unlikely to occur.

In scenarios characterised by low technical lifetimes and low levels of reuse—which against this background seem most likely—the closed-loop performance of the system becomes very dependent on two factors: the efficiency of metal recovery processes and the motivation level of consumers. This suggests an important role for recyclers such as the problem owner in promoting closed-loop performance, as well as a dependency on consumer behaviour. Specifically, it suggests that it may be useful for recyclers to engage in promoting flows of end-of-use phones from developing to industrialised countries, where they can be recycled more efficiently; and it suggests that recyclers may benefit from energising consumers to deposit their phones for collection.

The results presented above can form the basis for a set of policy recommendations. It is expected that implementation of these recommendations is not something that can be accomplished by any single actor. For instance, governmental regulation can play a role by mandating and incentivizing the collection of end-of-use phones, but the achievement of higher collection rates and metal recovery fractions also necessitate the participation of industry actors. The construction of this model has laid a foundation for engaging industry actors and policy makers in a dialogue en route to sensible policy measures. Further collaboration with the problem owner is in planning, in particular with the aim of constructing an extension of this model applicable specifically to the context of developing countries.

8.12 Conclusions

The purpose of developing this model was to identify leverage points for promoting a shift to closed-loop flow systems for metals in mobile phones. The model results underscore the point that there is no simple, silver bullet solution to achieving closed-loop performance of the studied system. As described in Step 8, single interventions carried out in isolation are limited in their potential to produce desired system performance. More promising is the implementation of multiple interventions in combination. As noted in the previous section, the most effective combinations of interventions appear to be those that simultaneously increase the consumer motivation level and the fraction of phones recycled by industrial metal recovery agents.

The model described in this chapter is an initial exploratory attempt at applying agent-based modelling to the domain of metal flows and electronic waste. Within this domain, agent-based modelling provides several advantages over more commonly applied techniques such as static substance flow analysis (Gordon et al. 2004; Johnson et al. 2005, 2006; Lifset et al. 2002; Mao et al. 2008) and dynamic equation-based techniques (Georgiadis and Besiou 2008; Kumar and Yamaoka 2007). In particular, agent-based modelling allows us the flexibility to represent societal metal flow systems in a more native manner, i.e. in a manner that conforms more naturally with such systems' real-world structures. In agent-based models, actors in the real-world system are represented as agents, allowing us to deliberately tailor their decision-making behaviour and track the evolution of their states over time and under different conditions. This added flexibility provides us with enhanced leverage to identify viable possibilities for steering the development of such systems.

Furthermore, the innovations introduced in this case—namely the representation of goods as discrete entities with unique properties and the introduction of continuous two-way communication between the model code and the ontology—serve to illustrate how state-of-the art techniques (see Chap. 9) can benefit the practice of agent-based modelling. These innovations, together with the code modules developed, lay a solid foundation for the development of future models within this domain. In particular, the setup of the model provides a generic structure for exploring the development of closed-loop flow systems for other types of materials and products. A model based on a similar structure is currently being developed to investigate the topic of informal electronics waste recycling in developing countries.

References

Bollinger, L. A. (2010). *Growing cradle-to-cradle metal flow systems: an application of agent-based modeling and system dynamics to the study of global flows of metals in mobile phones*. Master's thesis, TU Delft, Leiden University.

Ehrenfeld, J., & Gertler, N. (1997). Industrial ecology in practice: the evolution of interdependence at Kalundborg. *Journal of Industrial Ecology*, *1*(1), 67–79.

Frosch, R. A., & Gallopoulos, N. E. (1989). Strategies for manufacturing. *Scientific American*, *261*(3), 144–152.

Georgiadis, P., & Besiou, M. (2008). Sustainability in electrical and electronic equipment closed-loop supply chains: A system dynamics approach. *Journal of Cleaner Production*, *16*(15), 1665–1678.

Geyer, R., & Blass, V. D. (2009). The economics of cell phone reuse and recycling. *The International Journal of Advanced Manufacturing Technology*, *47*(5), 515–525.

Gordon, R. B., Bertram, M., & Graedel, T. E. (2006). Metal stocks and sustainability. *Proceedings of the National Academy of Sciences of the United States of America*, *103*, 1209–1214.

Gordon, R. B., Lifset, R. J., Bertram, M., Reck, B., Graedel, T. E., & Spatari, S. (2004). Where is all the zinc going: the stocks and flows project, part 2. *JOM Journal of the Minerals, Metals and Materials Society*, *56*(1), 24–29.

Johnson, J., Bertram, M., Henderson, K., Jirikowic, J., & Graedel, T. E. (2005). The contemporary Asian silver cycle: 1-year stocks and flows. *Journal of Material Cycles and Waste Management*, *7*(2), 93–103.

Johnson, J., Gordon, R. B., & Graedel, T. E. (2006). Silver cycles: the stocks and flows project, part 3. *JOM Journal of the Minerals, Metals and Materials Society*, *58*(2), 34–38.

Kumar, S., & Yamaoka, T. (2007). System dynamics study of the Japanese automotive industry closed loop supply chain. *Journal of Manufacturing Technology Management*, *18*(2), 115–138.

Lifset, R. J., Gordon, R. B., Graedel, T. E., Spatari, S., & Bertram, M. (2002). Where has all the copper gone: the stocks and flows project, part 1. *JOM Journal of the Minerals, Metals and Materials Society*, *54*(10), 21–26.

Lowe, E. A., & Evans, L. K. (1995). Industrial ecology and industrial ecosystems. *Journal of Cleaner Production*, *3*(1–2), 47–53.

Mao, J. S., Dong, J., & Graedel, T. E. (2008). The multilevel cycle of anthropogenic lead: II. Results and discussion. *Resources, Conservation and Recycling*, *52*(8–9), 1050–1057.

Mooallem, J. (2008). The afterlife of cellphones. *The New York Times*, pp. 38–43.

Rochat, D., Hagelueken, C., Keller, M., & Widmer, R. (2007). Optimal recycling for printed wiring boards in India. In *R'07 recovery of materials and energy for resource efficiency*.

Swiler, L. P. (2004). *A user's guide to Sandia's Latin hypercube sampling software*. Sandia National Laboratories.

Chapter 9
Next Steps in Modelling Socio-technical Systems: Towards Collaborative Modelling

A. Chmieliauskas, C.B. Davis, and L.A. Bollinger

Abstract In the practice of building models, we have encountered a number of methodological areas that need to be addressed in order to make the modelling process more robust, scalable, maintainable and transparent. In the context of modelling socio-technical systems (existing infrastructures and businesses) models are primarily data-driven: they contain heterogeneous agents, numerous assumptions and facts about agents and their environment. The data (the facts and assumptions used in the models) span multiple disciplines and are contributed by multiple domain experts. The focus of this chapter is to present new options for improving the management of model data. The issue of data management will be addressed at two levels: that of researchers who need to collaborate when creating and maintaining the model data, and that of data management within a simulation model. The methods proposed in this chapter rely on the use of Semantic Web technologies and philosophies to address data management issues. It is the goal of Semantic Web to make data understandable and useful for both machines and humans. In this chapter we will discuss the complications of modelling socio-technical systems and suggest uses of Semantic Web technologies to aid both collaboration between the modellers and knowledge management for the agents.

9.1 Complications of Modelling Socio-technical Systems

One of the challenges often encountered in developing models of complex socio-technical systems is that of having to deal with large amounts of diverse information. In other words, the problem is not just the volume of information but also the diversity of types of information. Different agents take their decisions based on various types of facts. These particular facts can describe diverse domains such as economics, policy and technology. Additionally, agents may have to consider both individual facts and aggregated information. For instance, an agent may need to assess individual facts such as finding the cheapest contract from someone with whom they have a high degree of trust. They also may want aggregated information that

A. Chmieliauskas (✉)
Delft, the Netherlands
e-mail: a.chmieliauskas@tudelft.nl

K.H. van Dam et al. (eds.), *Agent-Based Modelling of Socio-Technical Systems*, Agent-Based Social Systems 9,
DOI 10.1007/978-94-007-4933-7_9, © Springer Science+Business Media Dordrecht 2013

Fig. 9.1 Behaviour, query and data

is calculated using a large number of individual facts, such as the ratio of supply to demand for a particular commodity both produced and consumed in a simulated industrial cluster.

This ability of agents to *query* information contained in the model environment is a key requirement when building models of socio-technical systems. In order to explain the practical implications of this requirement we will approach the agent-based model from the standpoint of software development.

9.1.1 Agent-Based Model as Software

At a basic level, an agent-based model can be seen as a software application consisting of two main parts: data (e.g. system information decomposed into agents, facts and assumptions) and code that reads, writes and modifies that data. The data describes the state of the agents and their environment, while the behaviour of the agents is codified in queries into the data. The results of the queries are used in taking decisions (Fig. 9.1). Such abstract relationships between model components are similar to the standard paradigm used in artificial intelligence and robotics, where behaviour is composed of two types of components: acting and sensing (Brooks 1986). In agent-based models of socio-technical systems, sensing components are represented by the queries used by the agent to "sense" the environment and take decisions.

Contemporary agent-based model research advocates the use of ontologies for storing model data. Ontologies allow developers to encode model agents and assumptions via the use of ontology classes, properties and instances (van Dam 2009).

As an example of the types of the existing connections shown, consider the following: agents are connected to the technological objects they own; which in turn are connected to facts about their inputs and outputs; and these are connected to further information about the amounts, units and goods involved. Agents are also connected to other agents in networks comprised of economic and physical flows via contracts and trade. If one were to visualise this structure during the running of an agent-based model, it would appear as a giant, constantly evolving, simulated web of interconnected facts. Figure 9.2 depicts the ontology used for many of the models described in this book (see also Sect. 4.2) and gives an indication of what other ontologies might look like.

In models of socio-technical systems, agents must query this web in order to find out aspects of themselves and their simulated world. Whenever agents perform actions, they are also updating and adding new linkages in this web that will then in turn be queried by other agents as part of their own decision-making process. Thus management of large "graphs" of complex information, and enabling agents to make

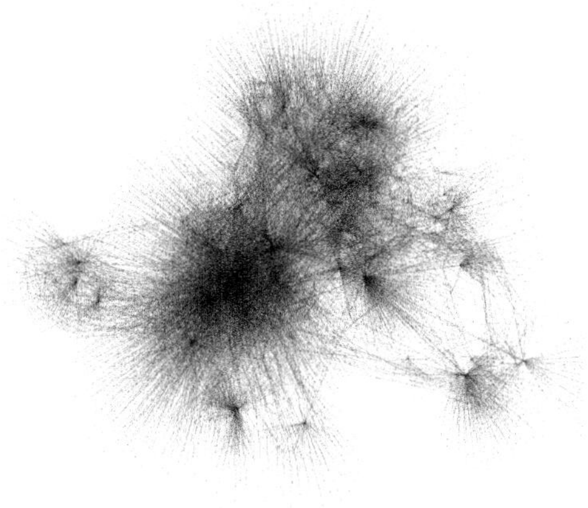

Fig. 9.2 Visualisation of the ontology used for many of the models described in this book. Nodes represent classes and instances, while edges represent the properties connecting them. The depiction employs a force-directed layout, meaning that nodes push away from each other, while edges act as springs and bring the nodes together. As a result, highly connected nodes migrate towards the center while less connected nodes are pushed to the outside

complex queries over this graph, is a requirement for agent-based modelling software functionality. As described further in this chapter, Semantic Web technologies provide software implementations that enable such functionality.

9.2 Applying Semantic Web Technologies and Methods for Modelling

The term "Semantic Web" is loosely defined as a set of philosophies and technologies that enable machines to understand the meaning of information on the World Wide Web (Antoniou and Harmelen 2008). While the Semantic Web vision is only just starting to materialise on the World Wide Web, the Semantic Web tools and practices can already be applied in managing data in other domains. The rest of this chapter will elaborate on the aspects of the Semantic Web technologies and methodologies that are relevant to modelling.

9.2.1 Semantic Web Technologies

The World Wide Web Consortium (W3C) has proposed a set of technologies for modelling, all intended to describe concepts and relationships between them. It includes, among others, the Resource Description Framework (RDF) (Klyne et al. 2004), RDF Schema (RDFS) (Brickley and Guha 2004) and the Web Ontology Language (OWL) (McGuinness and van Harmelen 2004). The data formats are intended

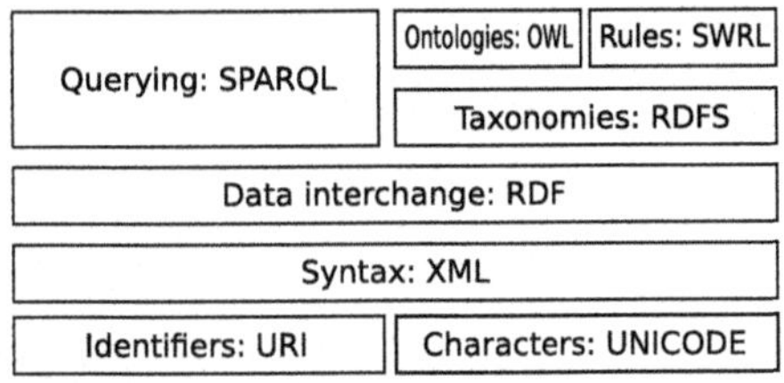

Fig. 9.3 The Semantic Web technology stack

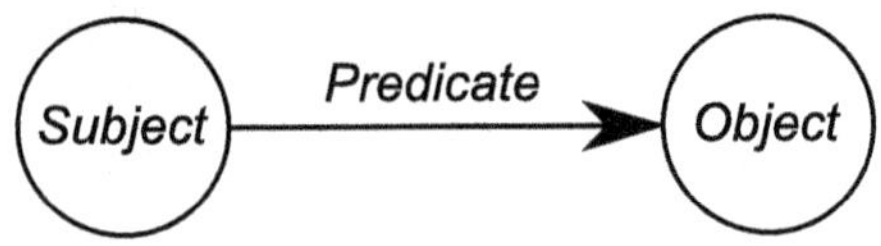

Fig. 9.4 A triple—the atom of the Semantic Web

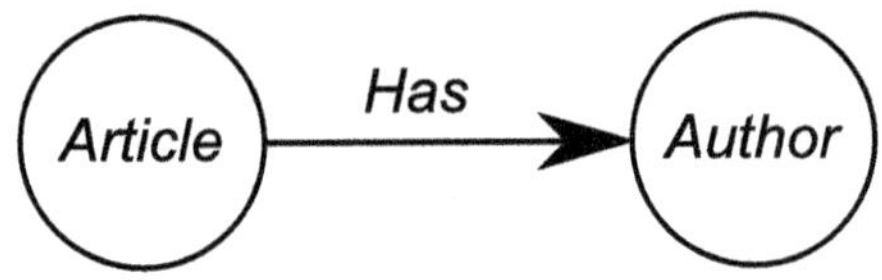

Fig. 9.5 RDF graph example

to be able to be read by both machines *and* humans. In the context of modelling, the Semantic Web technologies allow the formalisation of knowledge in formats that are accessible to both modellers and agents in the model. OWL is the Semantic Web standard to encode domain ontologies. It is also a subset of RDF, the native tongue of all Semantic Web technologies (see Fig. 9.3).

At the core of RDF is a *triple*: an expression of a subject, a predicate and an object (see Fig. 9.4). Everything in RDF is expressed in triples that encode statements about concepts and the relationships between them. For example, this article (subject) has (predicate) an author (object) (see Fig. 9.5).

Such triples combine to create a graph of data that has no predetermined boundaries. The graph is also not limited by any predetermined structure beyond the structure of a triple. This enables a crucial feature of the RDF data format, namely that the creator of the graph is not required to know the resulting data structure before starting to add information to it. The triple format allows for utmost flexibility, which is not the case in relational or object-based data structures (Bergman 2009). At the same time, the format can be interpreted by a computer, which is not the case with raw data in plain text.

The flexibility of the triple allows for easy encoding of the model facts, updating, reusing and sharing the data without additional overhead. When using a stricter data format (for example: a relational, object or frame-based ontology), any modifications to the structure of the data create cascading effects for the whole data model. Effectively, the triple format is the single most important factor in enabling encoding data collaboratively and without a central authority. This is further enabled due to the fact that RDF uses Universal Resource Indicators (URIs) to provide unique identifiers to objects in the data being described. An analogy to this would be the primary key used within a database. Data sets can be distributed across the web, in a similar fashion to how web pages already link to each other. This is radically different from any other more structured approach (be it a relational database or an ontology) that imposes a top-down design on the data being encoded.

From this, we can see that RDF is a very flexible means of describing complex data. One is not limited by a hierarchical framework but is free to represent knowledge in the form of a property graph. While this provides a means of storing data, what is still needed is a way of navigating and retrieving information from this graph; this will be discussed next.

9.2.2 Using SPARQL to Extract Structured Data

SPARQL (pronounced 'sparkle') is an RDF query language that allows querying of the graph by specifying the desired data patterns (Prud'hommeaux and Seaborne 2008; SPARQL query language for RDF 2008). The desired data patterns are specified in triples. Remembering the example data graph in Fig. 9.5, the SPARQL query in Algorithm 9.1 would search for the author of the article:

Algorithm 9.1 Example SPARQL query for the author of an article

```
select ?author where {
    :Article :Has ?author
}
```

The query follows the triple structure. ":Author" is the subject, ":Has" is the predicate, and the variable (unknown) "?author" is the object. In this case the variable is the object, but it could be the subject or the predicate, any combination or all three. So the query in Algorithm 9.2 would return every triple stored in the RDF store, meaning that it would retrieve every known statement about everything.

Algorithm 9.2 SPARQL query for all known triples

```
select ?subject ?predicate ?object where {
    ?subject ?predicate ?object
}
```

Without going into further detail of SPARQL syntax, it is sufficient to say that by using SPARQL functionality the modeller can specify the intended data pattern and extract the required data in the structure required by the model. In other words,

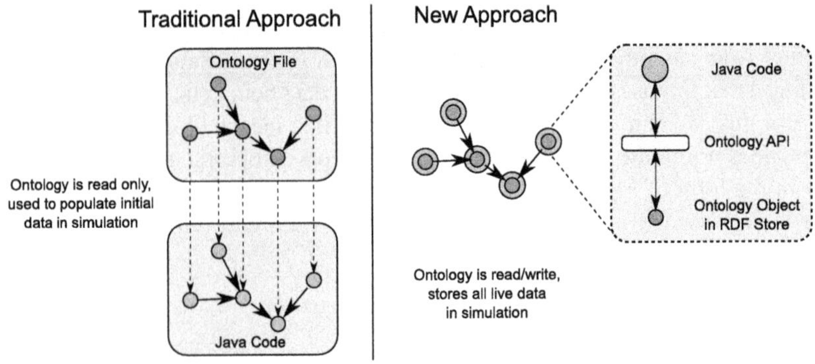

Fig. 9.6 The traditional and proposed approach to agent-based modelling with ontologies

SPARQL allows the modeller to lay custom structure (ontology) over the loose RDF data. The benefit of SPARQL over the ontology method lies within the decoupling of the data and their structure. The data are loosely structured and can be shared and extended easily, simply by adding arbitrary triples to it. The data structure is encoded within the SPARQL queries and is part of the specific model at hand. The queries are custom views of the data required by the model logic and the underlying research approach.

9.2.3 Using SPARQL Within Simulations

Used in simulations, SPARQL can be advantageous as a method for agents to collect information about their environment and to debug the simulations themselves. If one wishes to use SPARQL within simulations, this means that all the data used within a simulation must be kept within an RDF data store that is accessible through a SPARQL query engine.

The case study defined in the next section used Jena,[1] a Semantic Web framework written in Java which allows for both the storage and querying of RDF data. For this case, the simulation code used only defined behaviour of the agents and accessed all data through an application programming interface (API) provided by Jena, as illustrated in Fig. 9.6. To avoid confusion, when the word "ontology" is used in the text below, it refers to the set of data stored within the RDF data store. The ontology API refers to the Jena API which allows one to interact with the RDF data store.

For the previous models described in this book, with the exception of the model in Chap. 8, all data about the simulation was stored within POJOs (a common abbreviation for "Plain Old Java Objects", in other words, "normal" simple java code) which are initialised from the ontology at the start of a simulation. While this typically leads to code that executes more quickly than code that must retrieve data from

[1] http://jena.sourceforge.net/.

an RDF data store, this approach has a downside of making it difficult to manage information of a diverse nature. Part of the problem relates to what is called the *programming scope* of variables, i.e. the regions of the code where a variable is visible and can be read. For example, within Java, a variable defined within a *for* loop will not be accessible outside of it, and variables defined within a class are only visible outside of that class if it is declared as *public*. For everything that agents need to have knowledge about, one needs to have a way to make this information visible to them. If agents are to be omniscient, then essentially all variables must be globally visible. Using global variables has been considered an unproductive software development practice, making the program logic difficult to understand and maintain (Wulf and Shaw 1973).

In the past we have made information accessible by maintaining lists of information that were contained in the main simulation class, to which all agents had access. The problem with this approach is the management and maintenance of such lists. This is especially true if parts of the code are worked on by many people and multiple methods are used that can affect the addition and deletion of members in the list. This situation is further complicated if one needs to piece together information contained within multiple lists. The essential problem is that information is being kept in two places, which is an artifact of programming limitations and not of the system being modelled. The simulation itself is essentially a giant property graph of many connected facts. From experience, we have seen that in performing simulations, people have the need to traverse this graph in surprisingly different ways to enable various types of agent reasoning. By storing data natively within an ontological structure that can be queried, agents have access to the giant property graph of facts that constitutes the latest state of the entire simulation.

9.2.3.1 SPARQL for Agent Intelligence

Another benefit of using SPARQL as a method to extract structured data is that it can be used as an interface for agents to collect information about the state of their environment. Making data structures understandable both to humans and machines is at the heart of the Semantic Web vision, and SPARQL enables that vision. Thus the queries used by the modeller to structure and extract data can also be used by agents. The SPARQL queries embody the researcher's view of the system. Agents using the same queries to gather intelligence for decisionmaking adhere to the same view of the system. The idea is again parallel to using the ontology as a common language between agents (and the researcher) in the system. It is not easy to explain the benefit of using SPARQL for agent intelligence and communication without elaborating on the functionality and the features of the language. However, for the purpose of this chapter it is sufficient to say that SPARQL offers the ability for agents to pose complex questions about the environment in much the same way humans do. While using an ontology brings structure to agent communication, using SPARQL brings better intelligence as it allows agents to perform more sophisticated analysis.

The case study presented in Sect. 9.3 demonstrates the suggested use of SPARQL queries for agent intelligence.

9.2.3.2 SPARQL for Simulation Debugging

In the situation of a model storing information in an RDF data store, it is possible to write to a file all of the data from the simulation, which can be immensely useful for debugging certain types of problems. To implement this, one would write a method that would check for any erroneous patterns in the data, and upon detecting something wrong, it would write the data to a file and stop the execution of the simulation. This file could then be opened by an ontology editor such as Protege 3[2] or Topbraid Composer, which could then be used to perform queries over the entire graph of facts that constitutes the simulation data.

This is a major departure from previous ways in which we have tried to debug simulations. One method would be to identify problem areas in the code and log the values for certain variables that have been encountered. This approach runs into the issue of the *programming scope* of variables mentioned above. An alternative is to use an Integrated Development Environment (IDE) such as Eclipse[3] and set breakpoints that will stop the execution of the simulation at specified points in the code and allow for examination of variable values. While this approach works well if one is checking the value of a single variable, it becomes incredibly inefficient when debugging a problem where an agent needs to reason over facts distributed across many objects. In this case, the modeller is presented with a graphical user interface and must track down all the variables that the agent is aware of and piece together how the agent is reasoning over them. Having a copy of the simulation data solves this problem, since the same queries used by the agents can also be used by modellers to understand the system. Since one has a snapshot of all the data in the simulation, in order to debug agent reasoning they would simply perform the same SPARQL query that the agent is performing. If this is not enough, then they can create their own custom queries to explore the dataset from different views and try to pinpoint the problem.

More creative ways of debugging may exist, and already we have experimented with techniques such as the storing of historical simulation data in an RDF structure. The downside of this approach is that some selection about what is stored may need to be done in order to prevent the amount of data in the RDF data store from becoming too large and slowing down the simulation too much. An example of where this approach was employed will be discussed further in the section on the case study.

---

[2]Protege 4 does not support SPARQL.

[3]http://www.eclipse.org.

9.2.3.3 Disadvantages

As mentioned before, this approach does not lead to efficient code in terms of execution speed. Repeated calls to an RDF data store API do slow down the code noticeably. Additionally, if one has a model that is largely equation-based, or does not involve diverse types of complex data, then this approach is not an appropriate strategy and may not offer large benefits.

9.2.4 Implementations by Other Researchers

What we propose is not completely a new approach. Kwak et al. (2009), Matheus et al. (2009), Zhou et al. (2010) have already used SPARQL as a means for managing information in simulations. Furthermore, Knublauch (2009) has demonstrated a simple computer game in which behaviour is defined by an ontology that is accessed through SPARQL by objects in the game. The key difference with what is proposed here is the application of Semantic Web technologies to complex socio-technical problems. The applications that we have found by other researchers typically cover a single domain, while the systems we model span multiple domains and use Semantic Web technologies as a means to managing large volumes of information of a diverse nature.

9.3 Case Study: Mobile Phone Recycling Networks

In the preceding sections of this chapter we described the benefits of using RDF and SPARQL as an alternative to shared ontologies, especially in the context of collaborative modelling. We have also demonstrated the full route of using the approach to extract, structure and map the data for use in a model. We will now demonstrate elements of this approach in more detail.

The first case study to employ this approach was the investigation of mobile phone recycling networks, described in Chap. 8. For the case of this particular model, in order to provide a more native representation of the system, it was desirable to represent mobile phones as discrete entities with unique properties that evolved over the course of a simulation. This was achieved by the representation of each mobile phone in the model as a unique instance in the ontology. When a mobile phone is first "manufactured" during simulation, an instance of that mobile phone is immediately created in the ontology. Each time this mobile phone instance and its properties are viewed or modified by an agent during the course of the simulation, a communication between the model code and the ontology is incited.

In addition to representing each mobile phone as a distinct entity, the historical status of all phones is also recorded, allowing one to follow their paths from manufacturing all the way until reuse, recycling, or disposal. This means that for each time step there is a record of the agent in possession of the phone, along with the phone's price and various properties such as the condition and age of the phone. Figure 9.7 shows how with this information, we can perform a query that returns a table listing the line-by-line history of every phone. However, this is not the only view on the data, and various slices can be made through it depending on the perspective in which we are interested. For example, one could perform a query to list all the distinct pathways between agents among which a phone is transferred throughout the entire simulation. In other words, this will show all the types of agents that receive phones from manufacturers, or the distinct types of agents that receive phones from collectors. This will clearly show in a summarised fashion whether the supply, recycling, and disposal chains self-assemble in expected ways. In practice, this has shown problems such as phones moving directly from the manufacturer to disposal due to phones not being sold and consequently exceeding their age limits. One could also perform queries that aggregate information to examine aspects of the total population of phones, such as grouping the phones by their condition and tallying the number that have each particular condition at every time step. Through this, one may notice trends such as the existence of large populations of new phones despite incentives for reuse.

This approach has also been used for much more complicated problems. In particular, we have created agent-based models that are mass balanced. During early simulation runs we noticed that mass was actually disappearing from the system. We suspected that this may have been due to the process in which a phone, at the end of its life, was converted back into its raw components, such as gold. In order to debug the problem, we constructed a complex SPARQL query that first found all the contracts that included something being traded between agents in the previous tick. Then, for all the agents which received a mobile phone in this transaction, it checked to see if this phone was not placed in the stock of the receiving agent. Based on how the simulation was set up, this implied that the phone was disassembled into pieces and not refurbished. Since the mobile phones were represented as distinct objects, this query checked to see if the exact phone from the transaction made it into the stock of the receiving agent. The implication of this is that even if the receiving agent received 100 mobile phones and lost only one, we would be able to easily find the error with this query. The query shown in Algorithm 9.3 was actually further expanded to list what the agent did have in stock, and by doing this we were able to notice that the problematic agent had actually received two phones but only had the equivalent amount of gold that would result from disassembling a single phone:

As shown, these types of queries can be incredibly powerful for debugging. It should be noted that they are also powerful in their flexibility, since new statements can be easily added to the query to make it more specific. Additionally, queries can be made as vague as necessary, which can also be very useful. This approach was used to debug an issue where we knew that an instance of type *UnitName* did not have a label, which caused an error in the code. With an initial query, we were able to locate the actual instance that was causing the problem, although this was not enough information to determine what the label should be named, since it was not clear what it was being used to describe. To fix this, we had to traverse a hierarchy of objects as shown in the query in Algorithm 9.4. This found an instance of

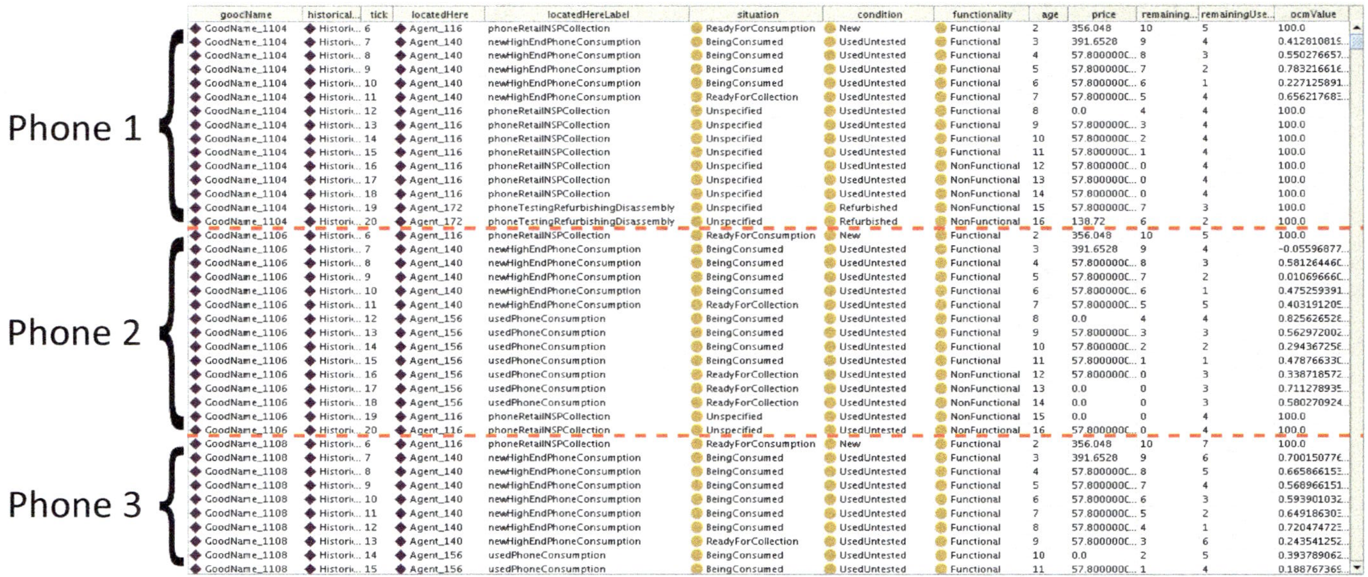

goodName	historical...	tick	locatedHere	locatedHereLabel	situation	condition	functionality	age	price	remaining...	remainingUse...	ocmValue
GoodName_1104	Histori...	6	Agent_116	phoneRetailNSPCollection	ReadyForConsumption	New	Functional	2	356.048	10	5	100.0
GoodName_1104	Histori...	7	Agent_140	newHighEndPhoneConsumption	BeingConsumed	UsedUntested	Functional	3	391.6528	9	4	0.412810819...
GoodName_1104	Histori...	8	Agent_140	newHighEndPhoneConsumption	BeingConsumed	UsedUntested	Functional	4	57.800000C...	8	3	0.550276657...
GoodName_1104	Histori...	9	Agent_140	newHighEndPhoneConsumption	BeingConsumed	UsedUntested	Functional	5	57.800000C...	7	2	0.783216616...
GoodName_1104	Histori...	10	Agent_140	newHighEndPhoneConsumption	BeingConsumed	UsedUntested	Functional	6	57.800000C...	6	1	0.227125891...
GoodName_1104	Histori...	11	Agent_140	newHighEndPhoneConsumption	ReadyForCollection	UsedUntested	Functional	7	57.800000C...	5	4	0.656217683...
GoodName_1104	Histori...	12	Agent_116	phoneRetailNSPCollection	Unspecified	UsedUntested	Functional	8	0.0	4	4	100.0
GoodName_1104	Histori...	13	Agent_116	phoneRetailNSPCollection	Unspecified	UsedUntested	Functional	9	57.800000C...	3	4	100.0
GoodName_1104	Histori...	14	Agent_116	phoneRetailNSPCollection	Unspecified	UsedUntested	Functional	10	57.800000C...	2	4	100.0
GoodName_1104	Histori...	15	Agent_116	phoneRetailNSPCollection	Unspecified	UsedUntested	Functional	11	57.800000C...	1	4	100.0
GoodName_1104	Histori...	16	Agent_116	phoneRetailNSPCollection	Unspecified	UsedUntested	NonFunctional	12	57.800000C...	0	4	100.0
GoodName_1104	Histori...	17	Agent_116	phoneRetailNSPCollection	Unspecified	UsedUntested	NonFunctional	13	57.800000C...	0	4	100.0
GoodName_1104	Histori...	18	Agent_116	phoneRetailNSPCollection	Unspecified	UsedUntested	NonFunctional	14	57.800000C...	0	4	100.0
GoodName_1104	Histori...	19	Agent_172	phoneTestingRefurbishingDisassembly	Unspecified	Refurbished	NonFunctional	15	57.800000C...	7	3	100.0
GoodName_1104	Histori...	20	Agent_172	phoneTestingRefurbishingDisassembly	Unspecified	Refurbished	NonFunctional	16	138.72	6	2	100.0
GoodName_1106	Histori...	6	Agent_116	phoneRetailNSPCollection	ReadyForConsumption	New	Functional	2	356.048	10	5	100.0
GoodName_1106	Histori...	7	Agent_140	newHighEndPhoneConsumption	BeingConsumed	UsedUntested	Functional	3	391.6528	9	4	-0.05596877...
GoodName_1106	Histori...	8	Agent_140	newHighEndPhoneConsumption	BeingConsumed	UsedUntested	Functional	4	57.800000C...	8	3	0.58126446C...
GoodName_1106	Histori...	9	Agent_140	newHighEndPhoneConsumption	BeingConsumed	UsedUntested	Functional	5	57.800000C...	7	2	0.01069666C...
GoodName_1106	Histori...	10	Agent_140	newHighEndPhoneConsumption	BeingConsumed	UsedUntested	Functional	6	57.800000C...	6	1	0.475259391...
GoodName_1106	Histori...	11	Agent_140	newHighEndPhoneConsumption	ReadyForCollection	UsedUntested	Functional	7	57.800000C...	5	5	0.403191205...
GoodName_1106	Histori...	12	Agent_156	usedPhoneConsumption	BeingConsumed	UsedUntested	Functional	8	0.0	4	4	0.825626528...
GoodName_1106	Histori...	13	Agent_156	usedPhoneConsumption	BeingConsumed	UsedUntested	Functional	9	57.800000C...	3	3	0.562972002...
GoodName_1106	Histori...	14	Agent_156	usedPhoneConsumption	BeingConsumed	UsedUntested	Functional	10	57.800000C...	2	2	0.294367258...
GoodName_1106	Histori...	15	Agent_156	usedPhoneConsumption	BeingConsumed	UsedUntested	Functional	11	57.800000C...	1	1	0.47876633C...
GoodName_1106	Histori...	16	Agent_156	usedPhoneConsumption	ReadyForCollection	UsedUntested	NonFunctional	12	57.800000C...	0	3	0.338718572...
GoodName_1106	Histori...	17	Agent_156	usedPhoneConsumption	ReadyForCollection	UsedUntested	NonFunctional	13	0.0	0	3	0.711278935...
GoodName_1106	Histori...	18	Agent_156	usedPhoneConsumption	ReadyForCollection	UsedUntested	NonFunctional	14	0.0	0	3	0.580270924...
GoodName_1106	Histori...	19	Agent_116	phoneRetailNSPCollection	Unspecified	UsedUntested	NonFunctional	15	0.0	0	4	100.0
GoodName_1106	Histori...	20	Agent_116	phoneRetailNSPCollection	Unspecified	UsedUntested	NonFunctional	16	57.800000C...	0	4	100.0
GoodName_1108	Histori...	6	Agent_116	phoneRetailNSPCollection	ReadyForConsumption	New	Functional	2	356.048	10	7	100.0
GoodName_1108	Histori...	7	Agent_140	newHighEndPhoneConsumption	BeingConsumed	UsedUntested	Functional	3	391.6528	9	6	0.700150776...
GoodName_1108	Histori...	8	Agent_140	newHighEndPhoneConsumption	BeingConsumed	UsedUntested	Functional	4	57.800000C...	8	5	0.665866153...
GoodName_1108	Histori...	9	Agent_140	newHighEndPhoneConsumption	BeingConsumed	UsedUntested	Functional	5	57.800000C...	7	4	0.568966151...
GoodName_1108	Histori...	10	Agent_140	newHighEndPhoneConsumption	BeingConsumed	UsedUntested	Functional	6	57.800000C...	6	3	0.593901032...
GoodName_1108	Histori...	11	Agent_140	newHighEndPhoneConsumption	BeingConsumed	UsedUntested	Functional	7	57.800000C...	5	2	0.649186303...
GoodName_1108	Histori...	12	Agent_140	newHighEndPhoneConsumption	BeingConsumed	UsedUntested	Functional	8	57.800000C...	4	1	0.720474723...
GoodName_1108	Histori...	13	Agent_140	newHighEndPhoneConsumption	ReadyForCollection	UsedUntested	Functional	9	57.800000C...	3	6	0.243541252...
GoodName_1108	Histori...	14	Agent_156	usedPhoneConsumption	BeingConsumed	UsedUntested	Functional	10	0.0	2	5	0.393789062...
GoodName_1108	Histori...	15	Agent_156	usedPhoneConsumption	BeingConsumed	UsedUntested	Functional	11	57.800000C...	1	4	0.188767369...

Fig. 9.7 Output from a SPARQL query used for debugging a model run

Algorithm 9.3 Example SPARQL query used for debugging mass balance problems

```
SELECT * WHERE {
    #find a mobile phone
    ?goodName :value "genericMobilePhone" .

    #this phone is part of a "PhysicalFlow" object
    ?physcialFlow :goodName ?goodName .

    #This "PhysicalFlow" is a component of a "PhysicalFlowContract" object
    ?physcialFlowContract :physicalFlow ?physcialFlow .

    #the "PhysicalFlowContract" came from this agent
    ?physcialFlowContract :from ?fromAgent .

    #and went to this agent
    ?physcialFlowContract :to ?toAgent .
    ?toAgent :label ?agentName .

    #Find all the ?toAgent who received a physicalFlow (i.e. a mobile phone)
    #who have not actually placed the mobile phone in their stock
    #(i.e. the phone has gone missing and/or been transformed into something else)
    OPTIONAL{?to :hasOpOutputInStock ?inStockTuple .
    ?inStockTuple :goodName ?goodName . }
    FILTER(!bound(?inStockTuple))
}
```

something that was connected to another instance of something that was connected to this problematic instance of type *UnitName* without a label. Using this, we were able to determine that the problem occurred within the *ConstructionCost* specified for a particular instance of a *Technology*.

The use of RDF + SPARQL within an agent-based model can be a powerful enabler for more sophisticated agent queries, and also for helping the modeller to make sense of the data contained within the simulation. However, since RDF + SPARQL is very generic, as will be described in the next section, we believe that it can be usefully applied in other contexts that can aid the modelling process.

Algorithm 9.4 Example SPARQL query used for discovering a consistency problem in the knowledge base

```
select * where {
    #find something that is a type of UnitName
    #but does not have a label specified
    ?x rdf:type :UnitName .
    OPTIONAL {?x :label ?xlabel }.
    FILTER(!bound(?xlabel)) .

    #find what this UnitName instance is connected to
    ?c ?d ?x .

    #figure out what the instance above is connected
        to
    ?a ?b ?c .
} order by ?xlabel
```

9.4 Future Work: Enabling Collaboration Between Modellers

While RDF + SPARQL can aid in the process of running and debugging models, there are aspects of it that have larger implications which may be of interest to modellers. In particular, RDF is a format that is designed to allow for the interconnection of data across the World Wide Web, similar to how HTML documents can link to each other across completely different websites maintained by different groups. Just as the web is a distributed, decentralised body of interlinked knowledge, RDF allows for the creation of distributed, decentralised interlinked data sets. What this means is that it opens up opportunities for reducing the burden of data collection for the modeller, while allowing for different groups to maintain these datasets for their own interests.

9.4.1 Limitations of Using a Single Ontology for Modelling

Modellers face not just an issue of what to do with information once it is in the model, but also one of how relevant information is added to the model in the first place. As described in the preceding chapters, the use of an ontology has greatly enabled the extensibility of models and collaboration between researchers. However, we have found the use of a single ontology to be restrictive, too, as adapting the existing ontology to new uses creates significant overhead and even conflicts, rendering the existing models unusable. As a shared ontology used for multiple models grows, it can accumulate properties and classes to the point where it is difficult to comprehend and maintain.

In our experience, we have seen a "long tail" in terms of ontology reuse: a small number of elements, such as those defining the inputs and outputs of technologies, are re-used to a large extent (van Dam 2009), while the ontology also contains a large number of elements that may have been used for only a single case study. New modellers must invest a significant amount of time in order to understand the structure and to learn what is available and how this can be applied. The limited range of concepts that they will likely be using for a new case study must often be explained to them by somebody more experienced in using the ontology.

9.4.2 Enabling Multiple System Views

Paradoxically, the use of ontologies in modelling complex adaptive systems conflicts with one of the definitions of complexity. According to Mikulecky (2001), complexity is 'the property of a real world system that is manifest in the inability of any one formalism being adequate to capture all its properties'. While there are definite benefits of using ontologies in modelling socio-technical systems, the ontologies should not bind the modeller to one view or formalism of the system. Allowing multiple system views is critical in order to make models extensible and adaptable to new research questions and to allow for collaborative model building.

The question that will be discussed further is whether the use of multiple formalisms can be enabled without sacrificing the benefits of the ontology approach. In order to assess this, we find it useful to decompose a model into very generic building blocks: data, structure and behaviour. The distinction between the data and structure is easy to understand in the ontological terms of classes (structure) and instances (data). This decomposition is analogous to a popular software development pattern: model-view-controller (MVC) (Gamma et al. 1994).

In MVC terms the "model" is the data, and it specifies that multiple "views" can exist about the same data. The "controller" is responsible for the behaviour of the system. The goal of MVC is to decouple software components to reduce the complexity in architectural design and to increase flexibility and maintainability of code (Gamma et al. 1994). We are pursuing a similar goal in modelling.

Having decomposed a model into three generic blocks (see Fig. 9.8), it can be assumed that each of these blocks is suitable for sharing between models and reuse. However, practice shows that when modelling socio-technical systems, the level of abstraction and reuse varies between these blocks. Data (facts) are the most generic blocks that can be used in many models, while structure and behaviour are more specific and tailored to answer a definite research question. While this might not be the case in a formula-based, mathematical or behavioural agent-based model, it particularly applies to infrastructure models that are primarily data-driven.

This decomposition of a model suggests that in order to enable collaboration and the reuse of models it is more useful to focus on sharing the data rather than the structure or specific behaviour. We have to choose the lowest common denominator in order to be able to achieve the desired flexibility and scalability of our models.

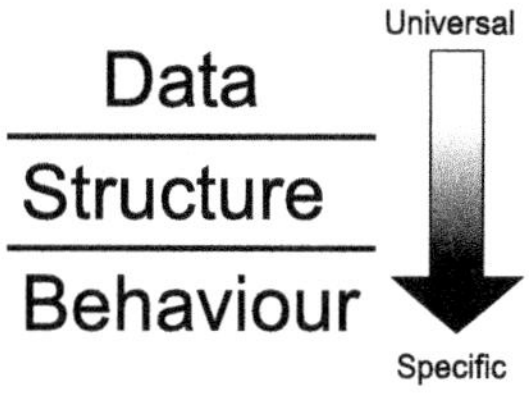

Fig. 9.8 Useful model decomposition following the MVC analogy

Nevertheless, we would like to retain the benefit of using the ontology as a formalism of any particular model and a common language between the agents in that model. A viable solution to this problem—sharing common data but having the ability to use a specific ontology in the model—can be achieved via the use of SPARQL query language.

9.4.3 Integrating Multiple Ontologies Together for Simulations

Using RDF for fact encoding and SPARQL to structure and extract the encoded data (RDF + SPARQL) presents a viable alternative to using a shared ontology. The main benefit of the RDF + SPARQL approach is that data can be created by many participants driven by different research agendas, without conflict or significant overhead (also called "creeping conceptualisation" (Allemang and Hendler 2008)), as is usually the case when using a single shared ontology. This allows for more efficient collaboration between researchers and enables the collection of data that can be shared between the different models. The approach also adheres to the principles of complexity. While the data is shared, numerous structures can be created for the same data, depending on the needs of a particular model or research question, representing a custom view of the system being modelled.

The custom SPARQL query results can then be used in the model, mapping them to the model objects in the same fashion as with the ontological instances. The following section will demonstrate the full RDF + SPARQL workflow in a simple example.

9.4.4 Simple Example of Integrating Multiple Ontologies

This section will demonstrate how one can create a custom remapped ontology from a subset of a larger ontology that may be available on the web. These ontologies may be maintained by other research organisations or government agencies, meaning that the burden of gathering and maintaining the data does not fall completely on the modeller. The point is not only that a single ontology can be remapped, but that this technique can be applied several times to different data sources. In other words, using Semantic Web, an ontology can be created for a model that draws together information from multiple sources. Some of the constructs and ideas might

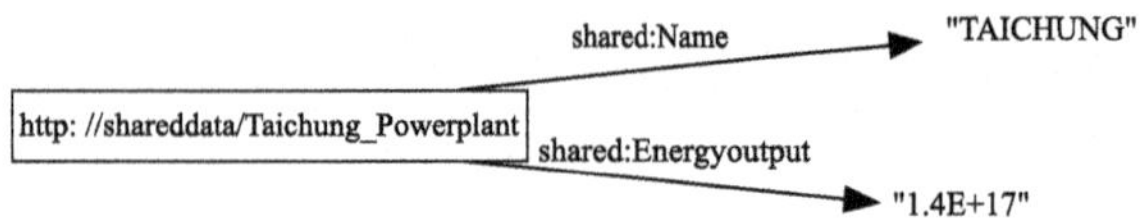

Fig. 9.9 Shared data graph

be confusing to the reader who is not technology-savvy. But to some extent, the purpose of this section is to stimulate interest in new enabling technologies.

Consider the graph in Fig. 9.9 within the pool of shared data.[4] The graph defines a power plant concept with two statements: "name" is "TAICHUNG" and "energyoutput" is "1.4E+17". In order to use the data in this graph the modeller has to transform it to the data structure required by the model. To transform the data we will use the SPARQL query in Algorithm 9.5.

Algorithm 9.5 Example SPARQL query to transform a data structure

```
1 PREFIX shared: <http://shareddata/>
2 PREFIX myOntology: <http://myontology/>
3 CONSTRUCT { #remap values from original data to new structure
4     myOntology:Taichung myOntology:title ?name .
5     myOntology:Taichung myOntology:power ?output .
6 } WHERE { #query to locate original data
7     shared:Taichung_Powerplant shared:Name ?name .
8     shared:Taichung_Powerplant shared:Energyoutput ?output .
9 }
```

There are three logical parts to the query. In the first part (lines 1–2) we define the namespaces for two data sets: the source and the result. The source is the shared data pool (http://shareddata/) and the result is the intended ontology for the model (http://myontology/). The second part of the query specifies the structure of the resulting data (lines 3–5). The third part specifies the triple pattern needed to retrieve data from the shared data pool (lines 6–9).

In effect, the results retrieved in the third part are aligned to the desired ontology specified in the second part of the query. Figure 9.10 shows the result of the query. This is a trivial example, and the only transformations to the graph are different namespaces and property names. The property "name" is transformed to "title", and the property "energyoutput" to "power". Following the same language constructs,

[4]Within a namespace http://shareddata/. A namespace is an abstract container or environment created to hold a logical grouping of unique identifiers or symbols (i.e. names), see http://en.wikipedia.org/wiki/Namespace_(computer_science).

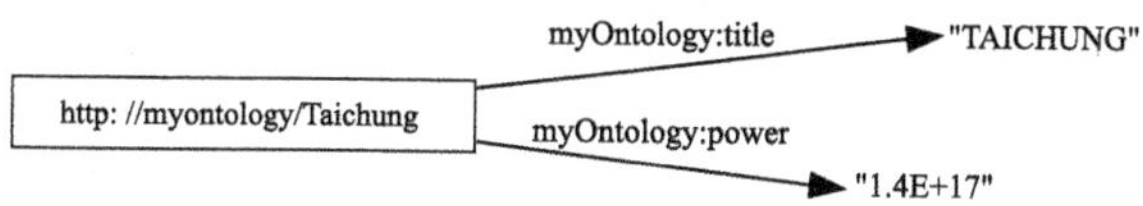

Fig. 9.10 Resulting data graph

complex queries can be written to redefine relationships between concepts and adapt to the necessary specification of the model.

The next step is to map the query result to a model object, for example as shown in Algorithm 9.6.

Algorithm 9.6 Example Java code of a Powerplant class including mapping from query results

```
1  public class Powerplant implements Powerplant {
2      String title;
3      Double power;
4
5      public Powerplant() {
6          // model logic here
7      }
8
9      public String getTitle() {
10         return title;
11     }
12     public void setTitle(String title) {
13         this.title = title;
14     }
15     public Double getPower() {
16         return power;
17     }
18     public void setPower(Double power) {
19         this.power = power;
20     }
21 }
```

The code above exemplifies the straightforward mapping between the model class (Powerplant) and query result that will be used populate the model object. The property "myOntology:name" (see Fig. 9.10) is mapped to class field "name" (line 2) and property "myOntology:power" to the field "power" (line 3).

In this simple example we have completed the full route of the RDF + SPARQL approach: extracting data from a shared data pool, aligning it to the model specification and mapping it to the object (agent) in the model. Such use of SPARQL enables the sharing of data between models and researchers, adapting it to fit the ontology of the model and avoiding conflicts that might arise from different views of the system. Furthermore, the RDF+SPARQL functionality is an enabling factor in researcher collaboration, creating complex models, connecting them and extending them in order to address new research questions in a more robust and timely fashion.

9.5 Conclusion

The use of Semantic Web technologies opens up several interesting opportunities for enhancing the model development process. Within simulation models, RDF allows for the creation of flexible information structures that can be easily extended and also integrated with data sets maintained by external organisations. SPARQL provides modellers with a powerful query language that can be used to create complex queries about what agents know about themselves and their simulated environment. The debugging of simulations may also be aided if modellers decide to store all live simulation data within an RDF, allowing them to use SPARQL to explore the intricacies of the entire set of data contained in the simulation. Particularly in models of complex socio-technical systems with large amounts of information of a diverse nature, this transparency can be quite an advantage.

The examples provided within this chapter represent existing trends occurring with the Semantic Web related to the availability of the software and data sets. They represent the initial steps we have taken in exploring their possible applications. While we have already gained benefits from our particular implementations, this technology is still maturing, and wider adoption will likely illuminate more disadvantages and opportunities when compared to other methods.

References

Allemang, D., & Hendler, J. A. (2008). *Semantic web for the working ontologist: modeling in RDF, RDFS and OWL*. San Mateo: Morgan Kaufmann.

Antoniou, G., & Harmelen, F. V. (2008). A semantic Web primer. In *Cooperative information systems* (2nd ed.). Cambridge: MIT Press.

Bergman, M. (2009). Advantages and myths of RDF. http://www.mkbergman.com/483/advantages-and-myths-of-rdf/.

Brickley, D., & Guha, R. (2004). RDF vocabulary description language 1.0: RDF schema. Online. http://www.w3.org/TR/rdf-schema/.

Brooks, R. (1986). A robust layered control system for a mobile robot. *IEEE Journal of Robotics and Automation*, *2*(1), 14–23.

Gamma, E., Helm, R., Johnson, R., & Vlissides, J. M. (1994). *Design patterns: elements of reusable object-oriented software*, 1 ed. Reading: Addison-Wesley.

Klyne, G., Carroll, J. J., & McBride, B. (2004). Resource description framework (RDF): concepts and abstract syntax. W3C recommendation 10. URL: http://www.w3.org/TR/rdf-concepts/.

Knublauch, H. (2009). SPIN box: A SPARQL-based computer game engine. URL: http://composing-the-semantic-web.blogspot.com/2009/01/spin-box-sparql-based-computer-game.html.

Kwak, J., Kang, Y., & Lee, Y. H. (2009). Implementation of a modeling and simulation tool for supporting RFID-based information system design. In *Proceedings of the Asia Pacific industrial engineering and management systems conference (APIEMS2009)*, Kitakyushu, Japan (pp. 1630–1636).

Matheus, C. J., Ulicny, B., Kokar, M. M., Baclaawski, K., Ng, W., Lau, L., Letkowski, J., Weyman, J., Dionne, R., & Parent, D. (2009). *Situational behavior modeling*. Technical report, Ft. Belvoir, Defense Technical Information Center. URL: http://stinet.dtic.mil/oai/oai?&verb=getRecord&metadataPrefix=html&identifier=ADA513027.

McGuinness, D. L., & van Harmelen, F. (2004). OWL web ontology language overview. Online. URL: http://www.w3.org/TR/owl-features/.

Mikulecky, D. C. (2001). The emergence of complexity: science coming of age or science growing old? *Computers and Chemistry*, *25*(4), 341–348.

Prud'hommeaux, E., & Seaborne, A. (2008). SPARQL query language for RDF. W3C recommendation. URL: http://www.w3.org/TR/rdf-sparql-query.

SPARQL query language for RDF (2008). W3C Recommendation. URL: http://www.w3.org/TR/rdf-sparql-query/.

van Dam, K. H. (2009). *Capturing socio-technical systems with agent-based modelling*. PhD thesis, Delft University of Technology, Delft, the Netherlands. ISBN 978-90-79787-12-8.

Wulf, W., & Shaw, M. (1973). Global variable considered harmful. *ACM SIGPLAN Notices*, *8*(2), 28–34.

Zhou, Q., Prasanna, V., Chang, H., & Wang, Y. (2010). A semantic framework for integrated modeling and simulation of transportation systems. *Relation*, *10*(1.116), 3458.

Index

A
Abnormal situation management, 152
Actor, 2, 4, 144, 146
Adaption, 30
Agent, 79
  properties, 57
  rules, 59
  state, 58
Agent-based model, 55, 57
  schedule, 66
Agent-based modelling, 5, 11, 54
  definition, 55
  history, 54
Algorithm, 89
Animation, 122
Artificial intelligence, 56, 246
Attractor, 31, 47, 118

B
Benchmarking, 167
BOINC, 112
Bottom-up, 55, 57, 75, 133, 179
Bounded rationality, 59
Building blocks, 145

C
Cap-and-trade, 203
CDU, 155
Cellular automata, 50, 54
Chaos, 45, 103, 107
Co-evolution, 27, 32, 45
Complex adaptive system, 1, 11, 44, 52
Complexity, 41
Context dependence, 61, 82
Coordinator, 12

D
Darwin, 27
Database, 114
Decentralised, 3
Decision rule, 59, 89, 148
Decision-making, 2, 6
  operational, 142, 152
  strategic, 142
  tactical, 142
Deterministic, 111
Distributed, 54
Distribution network operator, 2
Disturbance, 169
Documentation, 96
Dynamic, 6

E
Edison, 181
Einstein, 52
Emergence, 4, 22, 48, 51, 60, 75
Emissions trading scheme, 202
Environment, 4, 19, 26, 32, 61, 80
Equation-based model, 167
Evolution, 4, 5, 27, 133
Excel, 119
Experiment design, 105
Extreme values, 102, 168, 212

F
Feedback, 19
First-mover advantage, 67, 111
Fitness, 53
Fitness landscape, 33, 47
Formalism, 42, 52, 67
Full factorial, 108

K.H. van Dam et al. (eds.), *Agent-Based Modelling of Socio-Technical Systems*,
Agent-Based Social Systems 9,
DOI 10.1007/978-94-007-4933-7, © Springer Science+Business Media Dordrecht 2013

G
Game, *see* serious game
Game of life, 54
Generative science, 53, 75
Genetic algorithm, 60
GIS, 94
Git, 96
Greenhouse, 13, 76

H
Hierarchy, 2, 23, 42
High performance computing cluster, 113, 210
Hypothesis, 75, 105, 114, 123

I
In silico, 95, 105
In vitro, 74
Industrial ecology, 222
Industrial network, 152
Infrastructure
  electricity, 2, 201
  road, 3
  supply chain, 151
Interdependence, 17
Interdisciplinary, 42
Interoperability, 144
Intractable, 34, 133
Investment, 198, 203, 206, 225
Iteration, 79, 123, 131, 132, 135

K
Knowledge base, 147, 194
Kurtosis, 118

L
Latin hypercube sampling, 109, 114
Law of requisite variety, 52
Life-cycle, 221
Lighting, 181
  CFL, 182
  incandescent, 181
  LED, 182
Long tail, 258
Luminaire, 182

M
Market, 183, 201
  CO_2, 208
  crude oil, 165
  electricity, 208
  metals, 224
Matlab, 109, 119, 194
Meme, 28
Mental model, 134
Metal flow, 222
Mobile phone, 221
Model-view-controller, 258
Modelling
  agent-based models, 11, 54
  mathematical models, 53, 153, 242
  socio-technical systems, 7
  supply chains, 153
Monopoly, 201
Monte Carlo experiment, 110
Multi-agent system, 56
Multi-criteria analysis, 206
Multi-criteria decisions, 4, 59
Multi-disciplinary, 52
Multi-plant enterprise, 159
  actors, 160

N
Namespace, 260
Narrative, 88, 133
Nelder-Mead, 171
NetLogo, 54, 94
Network, 113, 122
  scale-free, 64, 191
  small-world, 64
Network (computing), 114
Neural network, 53, 60, 118
Next Generation Infrastructures, 7, 216

O
Object-oriented programming, 57
Observer-dependent, 21, 40
Occam's razor, 38
Octave, 119
Oil refinery, 154
  actors, 156
Ontology, 246
  definition, 84
  development, 145
  for mobile phones, 226
  for socio-technical systems, 144, 146, 160
  in agent-based modelling, 85
  tools, 84
Operation, 6, 142, 152
Operations research, 153
Optimise, 32, 171
Outlier, 103, 110, 118
OWL, 85, 247

P
Parallel, 88
Parameter sweep, 108, 194, 233
Path dependent, 28, 34
Pattern, 50, 117, 258

Policy, 4, 194, 241
Policy maker, 133, 203, 241
Power generation, 201
    actors, 204
Power law, 43
Power plant, 2, 182, 204, 260
Problem formulation, 74
Problem owner, 75, 132, 145
Procurement, 164
Protégé, 84, 210
Pseudo-code, 89, 101

Q
Query, 246

R
R (software), 109, 119, 125
Random number generator, 111
Random seed, 111
Randomness, 46, 109
RDF, 231, 247
RDF + SPARQL, 259
RDFS, 85, 247
Recycling, 222
    actors, 224
Reductionism, 23
Regulatory framework, 3
Repast, 54, 94, 111, 119, 210
Robotics, 246

S
Scenario, 105, 107
Self-organisation, 3, 50, 133
Semantic web, 231, 247
Serious game, 217
SETI@home, 112
Social network, 80
Socio-technical system, 1
    examples, 2
    power generation, 204
    supply chain, 153
Soup, 63
Space, 63
SPARQL, 231, 249
SPSS, 119
SQL, 114
Static substance flow analysis, 242
Statistics, 118
Subversion, *see* SVN
Supply chain, 151
    bull whip effect, 178
    lube oil, 160
    management, 178
    oil refinery, 155
SVN, 96, 98, 210
System, 14
    boundary, 18, 24, 77
    definition, 16
    general systems theory, 15
    hierarchy, 25
    history, 14
    level, 2, 3, 25, 51
    understanding, 4
System dynamics, 53, 130

T
Time, 42, 53, 62, 65, 78, 107
Top-down, 53, 133
Transition, 203, 215, 217
Transparency, 119, 262
Triple, 248

U
UML, 88, 91
Unit testing, 101, 232
Unstable, 107

V
Validation, 126
    face validation, 129, 240
    historic replay, 127, 197
    replication, 129, 177
Verification, 98, 103
Verification vs validation, 98, 103
Version control, 96, 146
Visualisation, 120, 131
    box plot, 195
VLCC, 155

W
Wiki, 43, 98

CPSIA information can be obtained
at www.ICGtesting.com
Printed in the USA
LVOW02*1716050317
526190LV00001B/86/P